# Communications
# in Computer and Information Science 406

Guilin Qi   Jie Tang   Jianfeng Du
Jeff Z. Pan   Yong Yu (Eds.)

# Linked Data and Knowledge Graph

7th Chinese Semantic Web Symposium and
2nd Chinese Web Science Conference, CSWS 2013
Shanghai, China, August 12-16, 2013
Revised Selected Papers

 Springer

Volume Editors

Guilin Qi
Southeast University, Jiangning, China
E-mail: qiguilin@googlemail.com

Jie Tang
Tsinghua University, Beijing, China
E-mail: jery.tang@gmail.com

Jianfeng Du
Guangdong University of Foreign Studies, China
E-mail: dududjf@gmail.com

Jeff Z. Pan
The University of Aberdeen, UK
E-mail: jeff.z.pan@abdn.ac.uk

Yong Yu
Shanghai Jiao Tong University, China
E-mail: yyu@apex.sjtu.edu.cn

ISSN 1865-0929 e-ISSN 1865-0937
ISBN 978-3-642-54024-0 e-ISBN 978-3-642-54025-7
DOI 10.1007/978-3-642-54025-7
Springer Heidelberg New York Dordrecht London

Library of Congress Control Number: 2013957362

CR Subject Classification (1998): I.2.4, I.2.6, F.4.1, C.2, H.5

*Typesetting:* Camera-ready by author, data conversion by Scientific Publishing Services, Chennai, India

Printed on acid-free paper

Springer is part of Springer Science+Business Media (www.springer.com)

# Preface

The Semantic Web is the next generation of the Web. It provides a common framework that allows data to be shared and reused across application, enterprise, and community boundaries. Semantic Web technologies facilitate building a large-scale Web of machine-readable and machine-understandable knowledge, and thus facilitate data reuse and integration so that the new generation of the Web can provide better applications and services. Nowadays, many scholars and experts are devoting themselves to applying the Semantic Web theories in specific practice, while in turn improving the Semantic Web standards and technologies according to the demand in practice. Web science involves the full scope of Web-related research and applications, and it integrates Web-related interdisciplinary research into a new field of scientific research. The jointly held Chinese Semantic Web Symposium and Chinese Web Science Conference aim to promote the expansion from the Semantic Web to Web science, and to discuss the core technologies of the next-generation Web, such as the Web and swarm intelligence, a new generation of semantic search, semantic and Web security, and so on.

Following the success of the 6th Chinese Semantic Web Symposium and the First Chinese Web Science Conference in Shenzhen, China, we organized the 7th Chinese Semantic Web Symposium and the Second Chinese Web Science Conference (CSWS 2013) in Shanghai, China. The theme of CSWS 2013 was "Linked Data and Knowledge Graph." A summer school was organized before the main conference. In the morning session of the summer school, two experts in Linked Open Data, Dr. Victor de Boer and Dr. Knud Möller, gave lectures on linked data principles and practice. In the afternoon session of the summer school, experts from three well-known IT companies (Google, Baidu, Sougou) introduced their work on the knowledge graph. Dr. Victor de Boer also gave an invited talk at the main conference. The title of the talk was "Linked Data for Digital Heritage and History." The other two invited speakers were Dr. Hua Wu from Baidu and Dr. Jeff Z. Pan from the University of Aberdeen. The title of Dr. Hua Wu's talk was "Internet-Based Semantic Analysis Technologies and Applications" and the title of Dr. Jeff Z. Pan's talk was "Handling Data Variety: The Semantic Web Approach."

This volume contains the papers presented at CSWS 2013. The conference received 48 submissions. Each submission was assigned to at least three Program Committee (PC) members to review. After a rigorous reviewing process, 14 research papers were selected for publication as full papers in the proceedings, five research papers were selected for publication in the *Journal of Southeast University* (Chinese Edition), and seven papers were accepted as short papers. The poster and demo track received five submissions, of which three were accepted.

These submissions cover a wide range of topics, including semantic search, ontology reasoning, social Semantic Web, knowledge graph, etc.

We would like to thank the excellent work of the PC. Each PC member was assigned three to four papers to review and their timely and professional reviews were helpful for us to select submissions with high quality. We also thank everyone who was involved in organizing the conference, especially Prof. Juanzi Li, Prof. Zhiqiang Gao, Prof. Huajun Chen, Prof. Dongyan Zhao, Dr. Haofen Wang, Dr. Wei Hu, Dr. Jie Bao, and several master students from Apex lab.

November 2013                                                       Guilin Qi
                                                                     Jie Tang
                                                                  Jianfeng Du
                                                                  Jeff Z. Pan
                                                                     Yong Yu

# Organization Committee

## General Chair

Yong Yu               Shanghai Jiao Tong University
Guilin Qi              Southeast University

## Local Chair

Haofen Wang         Shanghai Jiao Tong University

## Program Chairs

Jie Tang               Tsinghua University

## Summer School Chairs

Junzi Li               Tsinghua University
Zhisheng Huang       Vrije University of Amsterdam

## Workshop chairs

Huajun Chen          ZheJiang University
Dongyan Zhao        Peking University

## Poster and Demos Chairs

Jeff Z. Pan            University of Aberdeen
Jianfeng Du           Guangdong University of Foreign Studies

## Sponsor Chairs

Wei Hu                Nanjing University

## Metadata Chairs

Jie Bao                     Samsung

## Proceedings Chair

Zhiqiang Gao                Southeast University
Huiying Li                  Southeast University
Haitao Zheng                Tsinghua University, Shengzhen

# Table of Contents

# A New Model to Compute Semantic Similarity from Multi-ontology

Lan Wang and Ming Chen

College of Information Technology, Shanghai Ocean University, Shanghai 201306, China
`wllps1988315@126.com`, `mchen@shfu.edu.cn`

**Abstract.** Many measuring semantic similarity, with different methods, are applied in Natural Language Processing, knowledge acquisition and information retrieval. Recently, some authors have extended some of the existing methodologies to support multiple ontologies to improve the correlation values. In this paper, a feature-based method with heuristic function is proposed to deal with multi-ontology. By comparing the correlation values attained in this method with those of Pedersen's biomedical benchmark, a higher accuracy is achieved.

**Keywords:** Semantic similarity, Feature-based, multi-ontology, correlation value, heuristics.

## 1    Introduction

Currently people can access and encounter overabundant textual electronic information of the Word Wide Web. Textual resources are hard to manage due to the lack of textual understanding capabilities of computerized systems. Estimation of the semantic similarity between words is of great importance in many applications dealing with textual data, such as Natural Language Processing (NLP) [1], knowledge acquisition [2] and information retrieval [3].

The growth of the so-called Information Society has produced an enormous amount of numerical data, which can be directly managed by means of mathematical operators; but the coherent interpretation of textual data is challenging [4]. The evaluation of the semantic similarity between words is one of the most basic tasks. Ontologies have been of great interest to the semantic similarity research community as they offer a structured and unambiguous representation of knowledge in the form of conceptualizations interconnected by means of semantic pointers. For the explicit knowledge structure provided, ontologies have been extensively used in compute similarity. Ontologies provide a formal specification of a shared conceptualization. WordNet is a general purpose thesaurus that describes and organizes more than 100,000 general English concepts, which are semantically structured in an ontological fashion. Since WordNet is for general purpose, not for special one, it has very little coverage in special domains, such as biomedical domain. To make the measures more effective in biomedical domain, Pedersen et al [5] built The Systematized Nomenclature of Medicine, Clinical Terms (SNOMED CT), The Mayo Clinic CORPUS OF clinical Notes and The Mayo Clinic Thesaurus and substituted the biomedical underlying knowledge sources. Classical ontology-based similarity

G. Qi et al. (Eds.): CSWS 2013, CCIS 406, pp. 1–10, 2013.

approaches, purely focused on the analysis of the hierarchical structure of ontology, are able to compute similarity in an efficient manner without depending on external resources and human supervision [6]. However, more than one ontology is not supported, and only detailed and taxonomically homogenous ontology is available. The exploitation of multiple input sources would perform better coverage and more robust similarity estimations.

## 2     Previous Researches

Many scholars, domestic or foreign, have researched semantic similarity measures which have traditionally been an interesting research area within the NLP field [7]. Ontology is a directed graph we can input terms to the concepts by mapping their textual labels. Classical Ontology-based similarity assessment usually falls into three types: Path-based measures [8], Feature-based measures [9], Measures based on Information Content [4]. Path-based measures are the most basic and simplest measures, by computing the minimum Path Length connecting their corresponding ontological nodes via is-a links [10]. Path-based measures only rely on all notions linked in the taxonomy to represent a uniform distance, and the evaluation of the graph model requires a low computational cost. However, because of only considering the shortest path between concepts pairs, several taxonomical paths are not taken into account, in addition, the number and distribution of common and non-common taxonomical ancestors are neither considered. Feature-based measures overcome the limitations of path-based measures, by not representing uniform distances but assessing similarity between concepts as a function of their properties. Tversky's model of similarity takes into account common and non-common features of compared terms [11]. Feature-based measures are limited in ontologies for dependency on weighting parameters to balance the contribution of each feature. Sanchez et al presented an ontology-based measure relying on the exploitation of taxonomical features [12]. Measures based on Information Content was proposed by Resnik [13], who stated that semantic similarity depends on the amount of shared information between two terms, a dimension which is represented by their Least Common subsumer (LCS) in ontology. Here there is obviously a problem that any pair of concepts with the same LCS will result in exactly the same semantic similarity. Lin's solution to this problem is to measure the likeness of the two evaluated concepts as the ratio between the amount of information needed to state their commonality and the information needed to fully describe them. [14].Jiang and Conrath's solution was to quantify the length of the taxonomical links to different the IC and subsumer of a concept [15]. Sanchez and Batet have applied the previous three classic IC-based similarity measures.

The above classical similarity approaches, in general, do not support more than one input ontology. The exploitation of multiple input sources would perform better coverage and more robust similarity estimations. The traditional methods map features between ontologies, or connect ontologies by creating a new imaginary root node, or create a new integrated ontology, or create a new larger ontology directly [11]. Rodriguez and Egenhofer proposed a method to determine the most similar (or equivalent) entity (concept) classes of a given concept in two ontologies, by matching common features, finding similar classes in ontologies, calculating precision and recall, and ranking the similarity [16] and go. Al-Mubaid et al proposed an ontology-structure-based technique for measuring semantic similarity across multiple ontologies in the biomedical domain within the framework of Unified Medical Language System (UNLS), by creating a new imaginary root node and then computing cross-modified path length between two concepts [17]. Sanchez et al analyzed the similarity between the modeled taxonomical knowledge and the structure of different ontologies [18]. They also identified different cases according to which ontology/ies input terms belong, proposed several heuristics to deal with each case, solved missing values when partial knowledge was available, and captured the strongest semantic evidence that result in the most accurate similarity assessment when dealing with overlapping knowledge by creating a new imaginary root node [19]. They also proposed extending IC-based similarity measures by considering multiple ontologies in an integrated way, and several strategies were proposed according to which ontology the evaluated terms belong [20].

## 3    Proposed Method

In this section, a new model of feature-based semantic similarity measure across Multi-ontology is presented. We follow a similar principle as proposed in the Tversky's model and present a heuristic to choose from which of these ontologies the similarity should be computed. This paper is going to combine feature-based semantic similarity measure and a heuristic which provides a high accuracy.

### 3.1    Similarity Measure

Edge-counting approaches are simple, they only rely on the graph mode of an input ontology whose evaluation requires a low computational cost but all links in the taxonomy must represent a uniform distance. Feature-based measures overcome the limitations of Edge-counting approaches regarding the fact that taxonomical links in an ontology do not necessary represent uniform distances. Feature-based measures rely on features like glosses or synsets which limit their applicability on ontologies and they are dependent on the weighting parameters that balance the contribution of each feature.

In this paper, we follow a similar principle as proposed in the Tversky's model which consider that the similarity between two concepts can be computed as a function their common and differential features. Tversky begin with a set of objects described as collections of features, and a similarity ordering which is assumed to satisfy the axioms of the Eq.(1) below, they derive a measure f on the feature space and prove that the similarity ordering of object pairs coincided with the ordering of their contrasts. A and B be the features of terms respectively, $A \cap B$ be the intersection between two sets of features of concept a, and A-B :$\{a \in A \wedge a \notin B\}$, B-A : $\{a \notin A \wedge a \in B\}$. $\theta, \alpha$ and $\beta$ are the weight parameters.

$$S(a,b) = \theta f(A \cap B) - \alpha f(A - B) - \beta f(B - A) \quad \theta, \alpha, \beta \geq 0 \quad (1)$$

In order to evaluate the amount of dissimilarity between concepts, we consider the features of the taxonomical categorization of concepts which is given by the ontology. Let us define a as the sets of features of the concepts, b as the sets of features of the concepts. Considered f(A) as the set of differential taxonomical features, f(B) as the set of differential taxonomical features. There has a concept c, we can define concepts between a and b as: $f(A) - f(B) : \{c \in A \wedge c \notin B\}$, in the same way, $f(B)-f(A) : \{c \notin A \wedge c \in B\}$. Using the same notation introduced by Tversky & Sanchez, measure was defined as:

$$Dis(a,b) = \log_2(1+ \frac{(f(A) \cup f(B))-(f(A) \cap f(B))}{|f(A)-f(B)|+|f(B)-f(A)|+|f(A) \cap f(B)|}) \quad (2)$$

Our feature-based measure overcomes the limitation that no tuning parameters will be used to weight the contribution of potentially scarce semantic feature which improve the generality of our measure.

## 3.2    Multi-ontology Similarity Assessment

Batet and Sanchez propose several heuristics to deal with different cases according to which ontology/ies input terms belong and restructuring with Edge counting approaches [19]. We estimate Multi-ontology similarity based on Batet and Sanchez's heuristics, the multi-ontology similarity scenario is simplified to three cases.

Case 1: Concepts C1 and C2 appear in only one ontology, the similarity is computed like a mono-ontology setting.

Example 1 let us consider the following portion of ontology (Fig.1). The set of features generated for the concepts L1 and L4 are: f (L1) = {L1, C7, C3, C0}, f (L4) = {L4, C7, C3, C0, C4}. The number of differential taxonomical features of L1 with respect to L4 is 1(i.e., L1) and the number of differential features of L4 is 2(i.e., L4, C4), and the set of common elements has a cardinality of 3(i.e., C7, C3, C0). We can calculate the dissimilarity between L1 and L4 applying Eq. (2), the dissimilarity between L1 and L4 is calculated as follows: $Dis(L_1, L_4) = log_2 \left(1+ \frac{6-3}{1+2+3}\right) = 0.58$

Case 2: Both concepts appear at the same time in more than one ontology. The similarity calculus is influenced by the different levels of detail or knowledge representation accuracy of each ontologies [16]. In this case, for combining the individual evidence, the overlapping knowledge should be aggregate. Ontologies are created by different experts about different areas. So the aggregation of different ontologies is challenging. The ambiguities and semantic inconsistencies are hard to deal with. A higher taxonomical detail or better knowledge representation accuracy, however, should be needed. Ontology knowledge is usually partial and incomplete, so common features of knowledge should be modeled through an explicit ontology engineering process. It would be desirable to be able to decide, according to a heuristic, which ontology provides the, apparently the minimum similarity score applying Eq. (3).

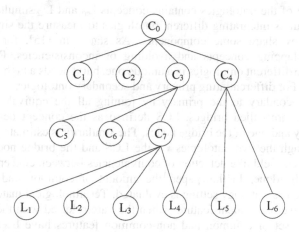

**Fig. 1.** Ontology example in WordNet

Example 2 concepts $L_{a1}$ and $L_{a2}$ belong to ontology $O_1$, concepts $L_{a1}$ and $L_{a2}$ belong to ontology $O_2$ (Fig.2). Applying the Eq. (2), in ontology $O_1$, Dis $(L_{a1}, L_{a2}) = 0.73679$ and in ontology $O_2$, Dis $(L_{a1}, L_{a2}) = 0.73696$, following the heuristic in Eq. (3), the final score is Dis $(L_{a1}, L_{a2}) = 0.73679$.

$$Dis(a,b) = \min \; dis(a,b) \qquad (3)$$

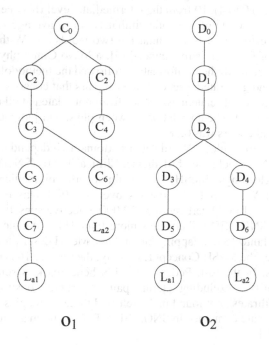

**Fig. 2.** Ontology o1 and o2

Case 3: None of the ontologies contains concepts $L_{a1}$ and $L_{a2}$ simultaneously. The current case requires integrating different ontologies to measure the similarity across ontologies if they share some components. As stated in [15], for dealing with ambiguous overlapping concepts and avoiding of inconsistencies, Rodriguez and Egenhofer merge different ontologies in a unique one, but it needs a high computational and human cost. For differentiating primary and secondary ontologies, Al-Mubaid et al connected the secondary to the primary by joining all the equivalent nodes. The equivalent nodes are called bridges. LCS defined as the concept belonging to the primary ontology and one of the bridge nodes. The similarity is estimated after the path is computed trough the two ontologies via the LCS and the bridge node [16]. In the ontology alignment field, the detection of equivalencies between concepts of different ontologies can be done. In this paper, the amount of common and non-common knowledge in a cross-ontology setting is evaluated. Terminological matching methods are used to discover equivalent features when they are referred with the same textual labels. Once the set of common and non-common features have been defined, the dissimilarity measure can be applied as presented in Eq (2).

## 4    Evaluations

The most commonly used benchmark is proposed by Miller and Charles [21], 38 undergraduate students were given 30 pairs of nouns from the original list of 65 studied by Rubenstein and Goodenough was used [22] and were asked to rate the similarity of each pair on a scale from 0 through 4, 10 were selected from the high level (between3 and 4 on a scale from 0 to 4), 10 from the intermediate level (between 1 and 3), and 10 from the low level (0 to 1) of semantic similarity . The average rating of each pair represents a good estimate of how similar the two words are. Within the biomedical field, benchmark of Pedersen is recommend [5], a Mayo Clinic physician (Alexander Ruggier, MD) trained in Medical Informatics, followed the methodology of Rubenstein and Goodenough and generated a set of 120 term pairs that consists of 30 pairs in each of four broad categories of relatedness values from not related at all (1) to very closely related (4) (see Table 1). The correlation between physician judgments was 0.68, and between the medical coders was 0.78.

In this paper, we select WordNet as domain independent ontology, and The Systematized Nomenclature of Medicine, Clinical Terms (SNOMED CT) and The Medical Subject Headings (MeSH) as domain-specific biomedical ontologies. The current version of WordNet (3.0) contains over 117,000 synsets, comprising over 81,000 noun synsets, 13,600 verb synsets, 19,000 adjective synsets, and 3,600 adverb synsets [23]. The SNOMED CT contains more than 311,000 concepts which unique meaning organized into 18 overlapping hierarchies. MeSH descriptors are organized in a tree which defines the MeSH Concept Hierarchy, the tree are 16 categories with more than 22,000 terms. In the test, Pedersen et al.'s benchmark is selected to Compute Semantic Similarity, excluding term pair "chronic obstructive pulmonary disease"-"lung infiltrates" not found in the selected three ontologies. For the remaining 29 pairs, all of them are contained in SNOMED CT, 25 of them are found in MeSH and 28 in WordNet [19].

**Table 1.** Test bed of 30 medical term pairs for benchmark of Pedersen [5]

| Term 1 | Term 2 | Physician | Coder |
|---|---|---|---|
| Renal failure | Kidney failure | 4.0 | 4.0 |
| Heart | Myocardium | 3.3 | 3.0 |
| Stroke | Infarct | 3.0 | 2.8 |
| Abortion | Miscarriage | 3.0 | 3.3 |
| Delusion | Schizophrenia | 3.0 | 2.2 |
| Congestive heart failure | Pulmonary edema | 3.0 | 1.4 |
| Metastasis | Adenocarcinoma | 2.7 | 1.8 |
| Calcification | Stenosis | 2.7 | 2.0 |
| Diarrhea | Stomach cramps | 2.3 | 1.3 |
| Mitral stenosis | Atrial fibrillation | 2.3 | 1.3 |
| Chronic obstructive pulmonary disease | Lung infiltrates | 2.3 | 1.9 |
| Rheumatoid arthritis | Lupus | 2.0 | 1.1 |
| Brain tumor | Intracranial hemorrhage | 2.0 | 1.3 |
| Carpel tunnel syndrome | Osteoarthritis | 2.0 | 1.1 |
| Diabetes mellitus | Hypertension | 2.0 | 1.0 |
| Acne | Syringe | 2.0 | 1.0 |
| Antibiotic | Allergy | 1.7 | 1.2 |
| Cortisone | Total knee replacement | 1.7 | 1.0 |
| Pulmonary embolus | Myocardial infarction | 1.7 | 1.2 |
| Pulmonary fibrosis | Lung cancer | 1.7 | 1.4 |
| Cholangiocarcinoma | Colonoscopy | 1.3 | 1.0 |
| Lymphoid hyperplasia | Laryngeal cancer | 1.3 | 1.0 |
| Multiple sclerosis | Psychosis | 1.0 | 1.0 |
| Appendicitis | Osteoporosis | 1.0 | 1.0 |
| Rectal polyp | Aorta | 1.0 | 1.0 |
| Xerostomia | Alcoholic cirrhosis | 1.0 | 1.0 |
| Peptic ulcer disease | Myopia | 1.0 | 1.0 |
| Depression | Cellulites | 1.0 | 1.0 |
| Varicose vein | Entire knee meniscus | 1.0 | 1.0 |
| Hyperlidpidemia | Metastasis | 1.0 | 1.0 |

The correlation values are obtained, as shown in Table 2. Pedersen et al.'s correlation values scored 0.588(Pedersen physicians) and 0.777( Perdersen coders) for SNOMED CT, 0.541 (Pedersen physicians) and 0.745( Perdersen coders) for WordNet, 0.588 (Pedersen physicians) and 0.782(Perdersen coders)for MeSH, 0.610 (Pedersen physicians) and 0.787(Perdersen coders)for SNOMED CT & WordNet, 0.580 (Pedersen physicians) and 0.775(Perdersen coders)for MeSH & WordNet, 0.615 (Pedersen physicians) and 0.798(Perdersen coders)for SNOMED CT & MeSH. In this test, correlation values scored 0.783 for SNOMED CT, 0.761for WordNet, 0.748 for MeSH, 0.788 for SNOMED CT & WordNet, 0.777 for MeSH & WordNet, 0.813 for SNOMED CT & MeSH. So, compared to Pedersen's, higher scores is gotten for SNOMED CT, WordNet, SNOMED CT & WordNet, MeSH & WordNet, SNOMED CT & MeSH while lower for MeSH. It can be generally concluded that a higher accuracy is achieved.

**Table 2.** Correlation values obtained by the benchmark of Perdersen

| Ontology | Pedersen physicians | Perdersen coders | Our method | Compared |
|---|---|---|---|---|
| SNOMED CT | 0.588 | 0.777 | 0.783 | higher |
| MeSH | 0.588 | 0.782 | 0.761 | lower |
| WordNet | 0.541 | 0.745 | 0.748 | higher |
| SNOMED CT & WordNet | 0.610 | 0.787 | 0.788 | higher |
| MeSH & WordNet | 0.580 | 0.775 | 0.777 | higher |
| SNOMED CT & MeSH | 0.615 | 0.798 | 0.813 | higher |

## 5    Conclusions

The Information Society has produced an enormous amount of textual electronic information. It is a fundamental and meaningful process to compute semantic similarity for natural language application, information retrieval and knowledge acquisition. A lot of researches have been developed for semantic similarity measures. Ontologies offer a structured and unambiguous representation of knowledge in the form of conceptualizations interconnected by means of semantic pointers and extensively used in compute similarity. Classical Ontology-based similarity assessment usually falls into three types: Path-based measures, Feature-based measures, Measures based on Information Content, proposed by Sanchez et al. But classical ontology-based similarity approaches, purely focused on the analysis of the hierarchical structure of ontology, are able to compute similarity in an efficient manner without depending on external resources and human supervision, but don't support more than one Ontology. The exploitation of multiple input sources would perform better coverage and more robust similarity estimations. The traditional methods match features between ontologies, or connect ontologies by creating a new imaginary root node, or create a new integrated ontology, or create a new larger ontology directly. Benchmark of

Pedersen is recommended within the biomedical field, and 30 term pairs were selected to compute similarity. In this paper, a feature-based approach is proposed which uses several heuristics to deal with three biomedical Ontologies, SNOMED CT, WordNet and MeSH, by matching features. The same 29 term pairs are used to compute similarity, excluding term pair "chronic obstructive pulmonary disease"-"lung infiltrates" not found in the selected three ontologies. Compared to Pedersen's, higher scores is gotten for SNOMED CT, WordNet, SNOMED CT & WordNet, MeSH & WordNet, SNOMED CT & MeSH while lower for MeSH. It can be generally concluded that a higher accuracy is achieved.

# References

1. Patwardhan, S., Banerjee, S., Pedersen, T.C.: Using measures of semantic relatedness for word sense disambiguation. In: Proceedings of the Fourth International Conference on Intelligent Text Processing and Computational Linguistics, Mexico City, pp. 241–257 (2003)
2. Chen, P., Lin, S.J., Chu, Y.C.: Using Google latent semantic distance to extract the most relevant information. J. Expert Systems with Applications 38, 7349–7358 (2011)
3. Hliaoutakis, A., Varelas, G., Voutsakis, E., Petrakis, E.G.M., Milios, E.E.: Information retrieval by semantic similarity. Int. J. Semantic Web Inf. Syst. 2(3), 55–73 (2006)
4. Sanchez, D., Batet, M.: A New Model to Compute the Information Content of Concepts from Taxonomic Knowledge. International Journal on Semantic Web and Information Systems 8(2), 34–50 (2012)
5. Pedersen, T., Pakhomov, S., Patwardhan, S., Chute, C.: Measures of semantic similarity and relatedness in the biomedical domain. Journal of Biomedical Informatics 40(3), 288–299 (2007)
6. Leacock, C., Chodorow, M.: Combining local context and WordNet similarity for word sense identification. In: WordNet: An Electronic Lexical Database, pp. 265–283. MIT Press, Cambridge (1998)
7. Java, A., Nirenburg, S., McShane, M., Finin, T.W., English, J., Joshi, A.: Using a natural language understanding system to generate semantic web content. International Journal on Semantic Web and Information Systems 3(4), 50–74 (2007)
8. Rada, R., Mili, H., Bichnell, E., Blettner, M.: Development and application of a metric on semantic nets. IEEE Transactions on Systems, Man, and Cybernetics 19(1), 17–30 (1989)
9. Petrakis, E.G.M., Varelas, G., Hliaoutakis, A., Raftopoulou, P.: X-similarity: computing semantic similarity between concepts from different ontologies. Digit. Inform. Manage. 4, 233–237 (2006)
10. Tversky, A.: Features of similarity. Psycological Review 84, 327–352 (1977)
11. Sánchez, D., Batet, M., Isern, D., Valls, A.: Ontology-based semantic similarity: A new feature-based approach. Expert Systems with Applications 39, 7718–7728 (2012)
12. Resnik, P.: Using information content to evaluate semantic similarity in a taxonomy. In: Proc. of 14th International Joint Conference on Artificial Intelligence, IJCAI 1995, pp. 448–453. Morgan Kaufmann Publishers Inc., Montreal (1995)
13. Lin, D.: An information-theoretic definition of similarity. In: Proc. of Fifteenth International Conference on Machine Learning, ICML 1998, pp. 296–304. Morgan Kaufmann, Madison (1998)

14. Jiang, J.J., Conrath, D.W.: Semantic similarity based on corpus statistics and lexical taxonomy. In: Proc. of International Conference on Research in Computational Linguistics, pp. 19–33. ROCLING X, Taipei (1997)
15. Rodríguez, M.A., Egenhofer, M.J.: Determining semantic similarity among entity classes from different ontologies. IEEE Trans. Knowl. Data Eng. 15(2), 442–456 (2003)
16. Al-Mubaid, H., Nguyen, H.A.: Measuring Semantic Similarity Between Biomedical Concepts Within Multiple Ontologies. IEEE Transactions on Systems, Man, and Cybernetics—Part C: Applications and Reviews 39, 389–398 (2009)
17. Sánchez, D., Solé-Ribalta, A., Batet, M., Serratosa, F.: Enabling semantic similarity estimation across multiple ontologies: an evaluation in the biomedical domain. Biomed. Inform. 45(1), 141–155 (2012)
18. Batet, M., Sánchez, D., Valls, A., Gibert, K.: Semantic similarity estimation from multiple ontologies. Applied Intelligenc. 38(1), 29–44 (2013)
19. Sánchez, D., Batet, M.: A semantic similarity method based on information content exploiting multiple ontologies. Expert Systems with Application 40, 1393–1399 (2013)
20. Miller, G.A., Charles, W.G.: Contextual correlates of semantic similarity. Lang. Cogn. Process. 6(1), 1–28 (1991)
21. Rubenstein, H., Goodenough, J.: Contextual correlates of synonymy. Commun. ACM 8(10), 627–633 (1965)
22. Fellbaum, C.: WordNet. Springer, Netherlands (2010)
23. Fanizzi, N., d'Amato, C., Esposito, F.: Inductive classification of semantically annotated resources through reduced Coulomb energy networks. International Journal on Semantic Web and Information Systems 5(4), 19–38 (2009)

# Linked Data Platform D2R$^+$

Qingling Chang[1,2], Shiting Xu[1], Yuanchun Zhou[1,*],
Jing Shao[1], Jianhui Li[1], and Baoping Yan[1]

[1] Computer Network Information Center, Chinese Academy of Sciences
[2] University of Chinese Academy of Sciences
{cql,xushiting,zyc,jingshao,lijh,ybp}@cnic.cn

**Abstract.** The concept of linked data has developed very fast since outlined in 2006 by Tim Berners-Lee, and now it has become a hot topic of web. It connects the traditional isolation data sets, and has become an important force to promote the development of semantic web. But firstly, we should publish the datasets into the web and interlink them with each other. On the other hand, the large number of data is heterogeneous, distributed, closed which bring in a serious impact on the sharing and use of data resources. How to expose them into the web under some unified standard or specification and providing meaningful association between them become the current problems to be solved. Thirdly, most of the current publishing tools do not support the interlinking of different datasets. In this paper, for relational database, we proposed the theory that relational database could be converted into ontology naturally, and based on which, we develop a linked data platform D2R$^+$ with the help of D2R open source. The platform could not only publish the data into web, but also support the interlinking among different datasets. What's more, the platform provides a user interface that allows the user to add data sets, as well as custom association relationship, and an intuitive way to browser the published data. It supports multiple types of data. Here, we take Qinghai Lake data and scientist data as an example to demonstrate the application of D2R$^+$.

**Keywords:** Linked Data, ontology, interlink different datasets.

## 1 Introduction

The central idea of Linked Data is to extend the web with a data commons by creating typed links between different data sources [1][2]. Technically, the term Linked Data refers to a set of best practices for publishing and connecting structured data on the web in a way that data is machine-readable, its meaning is explicitly defined, and it is linked to other external datasets, and can in turn be linked to from external datasets [2][3]. It has evolved as a powerful enabler for the transition of the document-oriented Web into the Semantic Web. It provides the tools and methods to share and expose metadata in a more unified and well

---

* Corresponding author.

G. Qi et al. (Eds.): CSWS 2013, CCIS 406, pp. 11–22, 2013.

interlinked manner, permitting both humans and machines to process web data [4]. The linked data has therefore evolved from a practical research idea into a very promising candidate for addressing one of the biggest challenges in the area of the Semantic Web vision: the exploitation of the Web as a platform for data and information integration [5].

On the other hand, the current data are mostly stored in heterogeneous, distributed, closed way, which impacts on the sharing of data sources seriously. Especially, the scientific data, most of them are got with high-end devices or in harsh environments, which are not reproduced. But they can not reflect their scientific value due to the isolation of these data, and the inherent knowledge cannot be tapped. For example, Chinese Academy of Science Database[1] and the Chinese Ecosystem Research Network Field Station ecological observation data sets[2], for the last 20 years, they have accumulated nearly 3 million paper data, 540,000 scientific data records, 6000 scientific data set. However, these data still provide services in an isolated state. So, how to publish and interlink them on the web is the important part of this study.

While there are more and more tools available for publishing Linked Data on the Web [3], there is still a lack of tools that support data publishers to interlink to other data sources. D2R$^+$ contributes to filling this gap. D2R$^+$ takes the ontology [6] concept as medium of data and RDF triples. It transforms the relational database to ontology first, and then publishes the ontology on the web with the Linked Data rules. It not only allows users to publish data, but also permits the publisher to interlink external data source. What's more, it provides a user interface and allows users to browse the published data intuitively.

This paper is structured as follows: Section 1 introduces the related work, which gives a summary of the popular linked data tools currently, and focuses on the D2R platform. In Section 2, we present the theoretical basis of the Linked Data. Based on the theory expounded in section 2, section 3 outlines the method of publishing and inter-linking of relational databases. We describe the architecture and function of the platform D2R$^+$ in Section 4. Section 5 concludes the article and Section 6 plans the future works.

## 2    Related Work

### 2.1    Publishing Tools

A variety of Linked Data publishing tools has been developed. The tools either serve the content of RDF stores as Linked Data on the Web or provide Linked Data views over non-RDF legacy data sources [2].

 * **D2R Server** D2R Server [7] is a tool for publishing non-RDF relational databases as Linked Data on the Web. Using a declarative mapping language, the data publisher defines a mapping between the relational schema of the

---

[1] http://www.csdb.cn
[2] http://www.cern.ac.cn

database and the target RDF vocabulary. Based on the mapping, D2R server publishes a Linked Data view over the database and allows clients to query the database via the SPARQL protocol.

* **Virtuoso Universal Server** The OpenLink Virtuoso server [8] provides for serving RDF data via a Linked Data interface and a SPARQL endpoint. RDF data can either be stored directly in Virtuoso or can be created on the fly from non-RDF relational databases based on a mapping.
* **Talis Platform** The Talis Platform [9] is delivered as Software as a Service accessed over HTTP, and provides native storage for RDF/Linked Data. Access rights permitting, the contents of each Talis Platform store are accessible via a SPARQL endpoint and a series of REST APIs that adhere to the Linked Data principles.
* **Pubby** The Pubby server [10] can be used as an extension to any RDF store that supports SPARQL. Pubby rewrites URI requests into SPARQL DESCRIBE queries against the underlying RDF store. Besides RDF, Pubby also provides a simple HTML view over the data store and takes care of handling 303 redirects and content negotiation between the two representations.
* **Triplify** The Triplify toolkit [11] supports developers in extending existing Web applications with Linked Data front-ends. Based on SQL query templates, Triplify serves a Linked Data and a JSON view over the application's database.
* **SparqPlug** SparqPlug [12] is a service that enables the extraction of Linked Data from legacy HTML documents on the Web that do not contain RDF data. The service operates by serialising the HTML DOM as RDF and allowing users to define SPARQL queries that transform elements of this into an RDF graph of their choice.
* **OAI2LOD Server** The OAI2LOD [13] is a Linked Data wrapper for document servers that support the Open Archives OAI-RMH protocol.
* **SIOC Exporters** The SIOC project [14] has developed Linked Data wrappers for several popular blogging engines, content management systems and discussion forums such as WordPress, Drupal, and phpBB.

## 2.2 The D2R Platform

Comparing with the variety of publishing tools, D2R is the most commonly used linked data tool.

The D2RQ [15] Platform is a system for accessing relational databases as virtual, read-only RDF graphs. It offers RDF-based access to the content of relational databases without having to replicate it into an RDF store. The D2RQ platform consists of D2RQ Mapping Language, D2RQ Engine and D2R Server. The architecture of D2RQ is shown in Fig.1.

The D2RQ Mapping Language [16] is a declarative language for describing the relation between a relational da-tabase schema and RDFS vocabularies or OWL ontologies. The D2RQ Engine is a plug-in for the Jena Semantic Web toolkit, which uses the mapping to rewrite Jena API calls to SQL queries against the database and passes query results up to the higher layers of the frameworks.

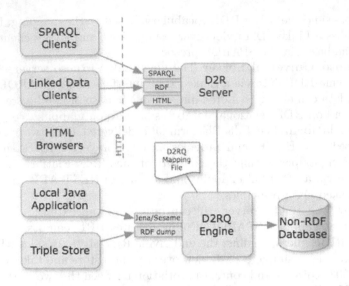

**Fig. 1.** D2R architecture

D2R Server is a tool for publishing relational databases on the Semantic Web. It enables RDF and HTML browsers to navigate the content of the database, and allows querying the database using the SPARQL query language.

Although D2R could publish relational database effectively, it does not support the interlinking to external data source. D2R$^+$ fills the gap with the help of D2R open source. What's more, it also provides an interface and allows users to browse data intuitively.

## 3    The Theory Base of Linked Data

### 3.1    The Four Principles

Berners-Lee (2006) outlined a set of 'rules' for publishing data on the Web in a way that all published data becomes part of a single global data space [1]:

1. Use URIs as names for things
2. Use HTTP URIs so that people can look up those names
3. When someone looks up a URI, provide useful information, using the standards (RDF, SPARQL)
4. Include links to other URIs, so that they can discover more things

Among the four rules, the first two principles are to establish the mechanism for standardized naming schema and the content of object; the third principle requires describing the content of object with structured and standardized way; and the forth principle asks to build links to other objects, which supports the associative retrieval from one object to the linked object. These four rules have become known as the 'Linked Data principles', and provide a basic recipe for publishing and connecting data using the infrastructure of the Web while adhering to its architecture and standards. All of them, the publishing is the most important node.

## 3.2  The Attentions of Publishing Linked Data

To publish Linked Data, we should take attentions on the below three issues [3]: How much data do you want to server? How often does your data change? How your data is currently stored. We can use the static RDF files to publish them if the data is little (hundreds of RDF triples or less), otherwise, we would better store the RDF triples in base, and select some server (such as Pubby ) to provide service for Linked Data. Or if the data changes constantly, then, it needs to bring in update mechanism or take the on-the-fly translation RDF.

Above all, the publishing of data involves a number of problems, especially, to publish huge amounts of RDF triples. Here, the platform D2R+ takes virtual RDF to realize the publishing of Linked Data.

## 4  The Publishing and Interlinking of Relational Database

There are various data storing forms, but the relational database is the most common traditional data stored way, furthermore, the other type data, such as pictures, video, audio, remote sense data, etc., in some extent, has the relationship with relational database. Because, the latent semantics on the web is indicated with their metadata, and the metadata are mostly stored in relational database. So the publishing and interlinking of relational database is particularly important. In view of this, the paper will focus on the relational database, and the interlinking with other type of data will be illustrated with a simple example.

### 4.1  The Basis of Relational Database to Ontology

Comparing the Ontology concept with the four rules of Linked Data, we will learn that there are many similarities between them. Such as, the unique identifier URI of Ontology is corresponding with the first two principles of Linked Data; as the web Ontology language recommendations of W3C, OWL is very similar to RDF, and OWL also supports SPARQL language, this characteristic is corresponding with the third principle of Linked Data; Ontology as the core of semantic web also interlinks the data, and which is corresponding with the forth principle. The correspondence between relational database and ontology is shown in table 1. So, we could transform the relational database to ontology first, and then publish the ontology according to the four principles on the web. And the idea is the main method to publish the traditional data on the web of this article.

**Table 1.** The correspondence between relational database and Ontology

| Relational database | Ontology |
|---|---|
| table | class |
| record | individual |
| field value | data attribute value |
| foreign key value | object attribute value |
| field name,foreign key,relationship table | attribute |

Furthermore, we express Ontology as RDF triple, then, what about the correspondence of relational database and RDF? In relational database, the data is stored as records, and we could represent a record with its unique key value. So, we could say that a record is multiple RDF triples when we take the key value as subject of RDF. The correspondence between relational database and RDF is shown in table 2.

In this table, we express both the key value in subject and object position, obviously, the key value on the object position indicates the foreign key value of one record. We also show the defined relationship which is built by user on the interface of D2R$^+$ platform.

**Table 2.** The correspondence between relational database and RDF

| RDF | Subject | Property | Object |
|---|---|---|---|
| | key value | column name | value |
| Relational | key value | foreign key | key value |
| database | key value | relationship table | key value |
| | key value | defined relationship | key value |

## 4.2   The Mapping Rules of Relational Database to Ontology

There are two way to transform relational database to ontology.

* virtual ontology. This way cannot support real-time very well. For all the users use one mapping file. It cannot update immediately when you add a new link, due to the numbers of online users. What's a worse, due to the virtual ontology, it has no copies, the security cannot be guaranteed and so on.
* real ontology. It is very difficult to build or update ontology data if the set is large, especially, when the data in relational database update frequently; and the real ontology makes the management of RDF inefficient; moreover, the spending to store RDF triples will expand.

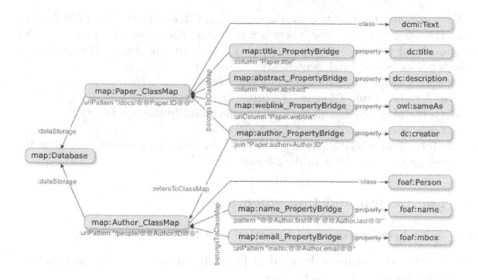

**Fig. 2.** The structure of an example D2RQ map

Generally speaking, the data scale is usually very large, to transform them to ontology one by one is unpractical, the usual practice is to make a set of grammar rules to do so. And the D2RQ Mapping language is a good demonstration. Fig. 2 shows the structure of an example D2RQ map.

The D2RQ mapping defines a virtual RDF graph that contains information from the database. This is similar to the concept of views in SQL, except that the virtual data structure is an RDF graph instead of a virtual relational table. The virtual RDF graph can be accessed in various ways, depending on what's offered by the implementation. Mappings can be written by hand in a text editor, but usually it is much faster to start with the generate-mapping tool that generates a skeleton "default mapping" from the database schema.

With the mapping language, users could map the table, column, row, column value and foreign key of relational database to class, attribute, resource, text and links of ontology.

## 4.3   The Interlinking of Different Databases

D2R can publish relational database efficiently, but it does not support the interlinking of different databases. It affects the data sharing and the realization of semantic web seriously. To make up the shortage, this article learns the open source of D2R, and realizes the function of interlinking to external database under the user interface help.

According to the characteristic of relational database, the central idea is to adopt multiple SQL sentences to make up the deficiency of not supporting retrieval in multiple databases about SQL.

For example, there are two databases D1 and D2, and the tables T1, T2 comes from D1, D2 separately, now, there is a field (e.g. Name) of T1 equals (or other relationships) to the field (e.g. Title) of T2, how to link them? The method in this paper as below: (SQL sentences)

*Select T1.name from T1 where T1.key='value',*

*Select T2.key from T2 where T2.title=T1.name*

With the above SQL sentences, the author modify the two function "de.fuberlin.wiwiss.d2rq.find .FindQuery" and "de.fuberlin.wiwiss.d2rq.sql.SQLIterator" in D2R open source to realize the interlinking between different data-bases.

## 5   The Design and Implementation

### 5.1   The Architecture and Function of D2R$^+$

The D2R$^+$ architecture is depicted in Fig.3. It consists of three models, 'user interface', 'server' and 'data browser'.

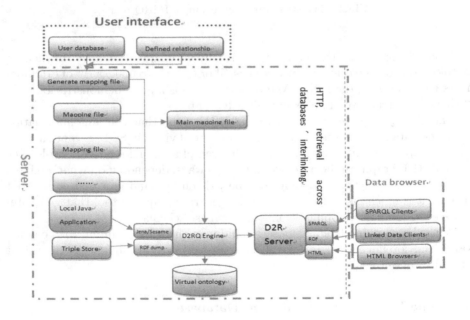

**Fig. 3.** D2R$^+$ architecture

The 'user interface' contains the 'user database' and the 'defined relationship'. The 'user database' allows users to select the dataset to publish, which overcomes the fixed pattern of general publishing tools to some extent. And the 'defined relationship' part allows users to define relationships between different tables or datasets, which greatly improves the flexibility and possibility of interlinking between different datasets.

The 'server' deals with the submitted databases and relationships and provides APIs for the browsers.

And the third model is 'data browser', which adds a vivid visualization browse way based on the D2R browse ways. After user submits his/her request, he/she could see the datasets and the all relationships among them on the web page. Especially, when the mouse move to some 'table' (in relational database), all the interlinked tables will be highlighted as 'red'.

## 5.2   The Implementation of D2R$^+$

The develop environment is: Windows xp, eclipse, jdk1.6 (or newer), tomcat6.0 (or newer) and D2R open source.

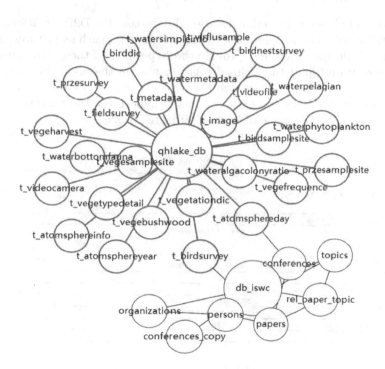

**Fig. 4.** Publishing and interlinking of 'qhlake' and 'iswc'

The 'user interface' is just a web page (.jsp), which provides a user entrance. When the user submits his dataset, the 'server' will generate the datasets into mapping files, if the D2R Engine calls the file, you could see the submitted dataset on the web page. If the user defines relationships, the process is the same with the submitted dataset. But if the relationship is between different datasets, the server will generate mapping files for every datasets and finally write all the mapping files into one main mapping file including the relationship mapping rules, when the D2R Engine calls the main mapping file, all the datasets and relationships will play on the web page.

The visualization data browser is realized With the D3.js technology, in which, we represent the table as 'circle', and the relationship between tables as 'line', and this also suits for different datasets, that is to say, the 'circle' could stand for the dataset and the 'line' could stand for the relationship between different datasets . The publishing and interlinking of Qinghai Lake survey data (qhlake) and scientist data (iswc) shows in Fig.4.

Obviously, from the figure, we could learn the relationship between scientific data and scientist clearly and intuitively, which may help us to find more useful information, and tag the potential knowledge.

### 5.3 Interlinking to Other Type Data

In addition to publish and interlink relational database, the D2R+ platform also support the functionality to interlink to other type data, such as pictures, audio and so on. To do this, users just provide the hyperlinks of these data, then, to define the relationship at the 'defined relationship' on the 'user interface' with the hyperlinks' and the links between them will shows on the web page (when the user look up the data and click the linked hyperlinks, the reference object will play). Fig.5 is an example of 'qhlake' links to 'youku'.

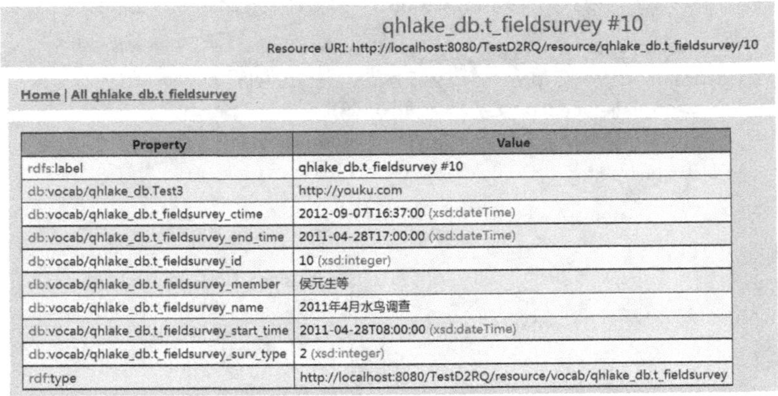
qhlake_db.t_fieldsurvey #10
Resource URI: http://localhost:8080/TestD2RQ/resource/qhlake_db.t_fieldsurvey/10

Home | All qhlake db.t fieldsurvey

| Property | Value |
|---|---|
| rdfs:label | qhlake_db.t_fieldsurvey #10 |
| db:vocab/qhlake_db.Test3 | http://youku.com |
| db:vocab/qhlake_db.t_fieldsurvey_ctime | 2012-09-07T16:37:00 (xsd:dateTime) |
| db:vocab/qhlake_db.t_fieldsurvey_end_time | 2011-04-28T17:00:00 (xsd:dateTime) |
| db:vocab/qhlake_db.t_fieldsurvey_id | 10 (xsd:integer) |
| db:vocab/qhlake_db.t_fieldsurvey_member | 侯元生等 |
| db:vocab/qhlake_db.t_fieldsurvey_name | 2011年4月水鸟调查 |
| db:vocab/qhlake_db.t_fieldsurvey_start_time | 2011-04-28T08:00:00 (xsd:dateTime) |
| db:vocab/qhlake_db.t_fieldsurvey_surv_type | 2 (xsd:integer) |
| rdf:type | http://localhost:8080/TestD2RQ/resource/vocab/qhlake_db.t_fieldsurvey |

**Fig. 5.** 'qhlake' links to 'youku' audio

From this perspective, we can say that $D2R^+$ provides us a semantic cloud service platform, because, if user wants to interlink the data, he just submits his data and defines the useful relationship, then the data and the relationship with semantic information will show on the web intuitively.

# 6 Conclusions

The article gives the theory base that is the relational database could transform to ontology naturally, and then discusses the rule of mapping file. With which, we develop the D2R+ platform under the open source of D2R help.

The D2R$^+$ platform expands the function of D2R, which not only publishes relational database, but also provides to interlink to external data source, also other type datasets. What's more, it provides a user interface, which allows users to submit dataset and define relationship between different sources. This makes the publishing and interlinking of data become more flexibility and practicality, and the interaction idea provides a new research model for the publishing and interlinking issues. Furthermore, the visualization browse way allows users to look up the data intuitively and vividly. Except for all the above mentioned, the platform also provides an easy and fast way to build ontology bases, whether the user is aware of the underlying structure of the data or not, which also provides a new idea for building ontologies.

Obviously, there are some shortages of D2R$^+$ platform, for example, the real-time and the security problems and so on.

# 7 Future Works

In future work, we plan to improve the real-time and security problems mentioned above first, then, we hope to realize the retrieval among multiple databases based on the D2R$^+$. Finally, we expect to study on link discovery framework for interdisciplinary research of integrated key area. And hope to publish the Qinghai Lake scientific data and discovery the links with the framework at last.

**Acknowledgements.** This work is partially supported by Natural Science Foundation of China under Grant No. 91224006 and 61003138, the Strategic Priority Research Program of the Chinese Academy of Sciences under Grant No. XDA06010202 and XDA05050601-4, the Special Project of Informatization of Chinese Academy of Sciences in the Twelfth Five-Year Plan under Grant No. XXH12504.

# References

1. Berners-Lee, T.: Linked data-design issues (2006),
   http://www.w3.org/DesignIssues/LinkedData.html (2011)
2. Bizer, C., Heath, T., Berners-Lee, T.: Linked data-the story so far. International Journal on Semantic Web and Information Systems (IJSWIS) 5(3), 1–22 (2009)
3. Mitsopoulou, E., Taibi, D., Giordano, D., Dietze, S., Yu, H.Q., Bamidis, P., Bratsas, C., Woodham, L.: Connecting medical educational resources to the linked data cloud: the meducator rdf schema, store and api. Proceedings of Linked Learning (2011)

4. Auer, S., Lehmann, J.: Creating knowledge out of interlinked data. Semantic Web 1(1), 97–104 (2010)
5. Bizer, C., Cyganiak, R., Heath, T.: How to publish linked data on the web. Retrieved 20, 2008 (2007)
6. Ontology(information science), http://en.wikipedia.org/wiki/Ontology_(information_science)
7. Bizer, C., Cyganiak, R.: D2r server-publishing relational databases on the semantic web. In: 5th International Semantic Web Conference, vol. 26 (2006)
8. Virtuoso rdf, http://www.openlinksw.com/dataspace/doc/dav/wiki/Main/VOSRDF
9. Strategies for building semantic web applications, http://notes.3kbo.com/talis
10. Cyganiak, R., Bizer, C.: Pubby-a linked data frontend for sparql, http://www4.wiwiss.Fu-berlin.de/pubby/
11. Auer, S., Dietzold, S., Lehmann, J., Hellmann, S., Aumueller, D.: Triplify: lightweight linked data publication from relational databases. In: Proceedings of the 18th International Conference on World Wide Web, pp. 621–630. ACM (2009)
12. Coetzee, P., Heath, T., Motta, E.: Sparqplug: Generating linked data from legacy html, sparql and the dom. In: LDOW (2008)
13. Haslhofer, B., Schandl, B.: The oai2lod server: Exposing oai-pmh metadata as linked data (2008)
14. Sioc exporters, http://sioc-project.org/exporters
15. D2rq-accessing relational databases as virtual rdf graphs, http://d2rq.org/
16. Sahoo, S.S., Halb, W., Hellmann, S., Idehen, K., Thibodeau Jr., T., Auer, S., Sequeda, J., Ezzat, A.: A survey of current approaches for mapping of relational databases to rdf. W3C RDB2RDF Incubator Group Report (2009)

# An Ontology-Based Approach
# to Extracting Semantic Relations from Descriptive Text

Da Huang and Wei Hu

State Key Laboratory for Novel Software Technology,
Nanjing Univiersity, China
dhuang.cn@gmail.com, whu@nju.edu.cn

**Abstract.** Linked Data have advantages over plain text, as data are organized in relations between information, which is convenient for learning and reasoning. However, most plain text with valuable information has not been converted into Linked Data form. Thus, we propose an ontology-based method to extract semantic relations from descriptive text about entities. Moreover, we conduct our experiment on the DBpedia dataset and design an automatic methodology to evaluate our ontology-based method as well as an intuitive method. As a result, we find out that our ontology-based method performs better than the intuitive one in general. At last, we analyze the results, and put forward our opinions on the difference between the two methods' performance.

## 1    Introduction

Since the concept of Semantic Web was proposed, Linked Data [1] have been more and more widely used in many areas. There are more and more people realizing that data are much more portable for learning and reasoning when organized in context-linked form rather than plain text. Therefore, Linked Data are considered to be easy-to-use knowledge, while the traditional plain text data are taken as hard-to-use knowledge.

However, plain text is still one of the most important forms of knowledge, and its amount is several times as much as Linked Data. Hence, we come up with the idea of extracting semantic relations from plain text. As most valuable knowledge is contained in descriptive plain text that is composed of sentences describing something specific, our job concentrates on extracting semantic relations from such kind of text. Fig. 1 shows an example of our job.

In this paper, we propose an ontology-based approach to figuring out the extraction task. Furthermore, we design an automatic experiment to evaluate our method and use it to compare ours with an intuitive method.

Our experiments and evaluations are based on the DBpedia dataset [2]. We choose DBpedia, because it is open-access, widely-used, high-quality, and most importantly, it provides us with some semantic relations which correspond to existing descriptive text on specific entities.

G. Qi et al. (Eds.): CSWS 2013, CCIS 406, pp. 23–36, 2013.
© Springer-Verlag Berlin Heidelberg 2013

This paper is organized as follows: Section 2 summarizes related work in this area. Section 3 describes an intuitive method and our ontology-based one. Section 4 talks about our experiment and evaluation on each method. Finally, Section 5 draws conclusions and discusses future work.

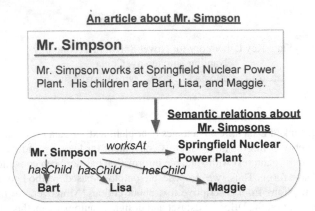

**Fig. 1.** An example of extracting semantic relations from descriptive text

## 2    Related Works

There have been lots of works concerned with extracting semantic relations from plain text.

P. Buitelaar et al. have made an overview about ontology learning from text [3]. They pointed out that semantic relations can be extracted on the several levels, including the concepts level, concepts hierarchy level and the relations level. The first two can be taken as schema level, while the last one as instance level. As for the former one, M. Vela [4], P. Cimiano [5] and their teammates have designed methodologies for it respectively. Considering our work lays on the latter one, we would not describe their works in detail here.

W. Wong et al. have summarized existing techniques in the extraction task mainly at the instance level [6]. Several techniques are included in following works.

L.K. McDowell and M. Cafarella have built an ontology-driven information extraction (IE) system named OntoSyphon [7]. Their IE system was able to extract instances of relationship *"class"* with the help of the ontology's class hierarchy.

K. Khelif et al. have put forward an ontology-based method to support text mining and information retrieval in the biological domain [8]. Unlike the last work, their approach could extract instances of any relationship from plain text by using text patterns. However, these patterns came from relationships definitions and biologists' suggestions, but not generated in an automatic way.

E. Agichtein and L. Gravano have created a system called Snowball, which could extract relations from large plain-text collections [9]. Its architecture was first extracting patterns and then matching text with patterns, which was similar to ours. However, our approach pays more attention to the syntactical structure of text, which helps patterns fit better.

K. Fundel et al. have come up with a method named RelEx, which extracted relations by using dependency parse trees [10]. The method did not relay on the ontology, but on dependency parse trees. In this paper, we put forward an intuitive method based on theirs, and compared our ontology-based method with it.

The approach to extracting semantic relations that we proposed in this paper is different from all mentioned above. Firstly, our method learns accurate and flexible templates from the corresponding relations between structural data and some plain texts. Secondly, some techniques of syntactical analysis are absorbed in our approach. Last but not least, our experiments and evaluations are conducted on the DBpedia dataset, which is much closer to the real situation of application, as data in the DBpedia are derived from Wikipedia.[1]

# 3     Methodology

Before putting forward any specific method, we have to consider our motivation in a more rigorous way. Definition 1 shows the formalized definition of our motivation.

**Definition 1.** Take all descriptive text as $\mathbb{T}$ and the current ontology as $\mathbb{O}$. There are infinite but countable binary relationships, which correspond respectively to $p_1$, $p_2$, $\dots$, $p_n$, $\dots$ Among them, $p_i(i \in \mathbb{N}^+)$ is a name that can express the semantic meaning of the $i$-th relationship. The relation universal set of each relationship is denoted respectively by $R_1$, $R_2$, $\dots$, $R_n$, $\dots$ Then, "Extracting Relations from Descriptive Text" can be defined as the function $f: P(\mathbb{T}) \times \{\mathbb{O}\} \to P(\bigcup_{i \in \mathbb{N}^+} R_i)$.[2,3] If $f(t, \mathbb{O}) = \bigcup_{i \in \mathbb{N}^+} r_i$ ($t \subseteq \mathbb{T}, r_i \subseteq R_i$), then for any $\langle s, o \rangle \in r_i$ ($i \in \mathbb{N}^+$), we can find $s, o \in t$, $p_i \in \mathbb{O}$ and some $c \in t[s, o]$ with $p_i \sim c - s - o$. Here, $\sim$ stands for semantic equivalence, while $t[s, o]$ denotes the set of sub-contexts induced by $s$ and $o$ in $t$.[4]

In this section, we describe two methods for semantic relations extraction according to Definition 1. One is an intuitive method taken as baseline, while the other one is an ontology-based approach we propose.

## 3.1     An Intuitive Method

Firstly, we show an intuitive method to extract semantic relations, which is a straight realization of Definition 1. This method is based on the RelEx designed by K. Fundel and his co-workers [10], as mentioned in last section. In Section 4, we will compare this method with ours.

As Fig. 2 shows, the method is composed of two steps. The first step is to generate a dependency graph for the original text, which is done by using a tool named Stanford CoreNLP.[5] The second one is to extract relations based on the dependency graph. Here, we just describe the second step.

---

[1] http://www.wikipedia.org/
[2] Here, $P$ is the symbol of power set. $P(S) = \{s | s \subseteq S\}$.
[3] $\bigcup_{i \in \mathbb{N}^+} S_i = S_1 \cup S_2 \cup \dots \cup S_n \cup \dots$
[4] $t[s, o]$ is set, because $s$ and $o$ may be mentioned more than once in $t$.
[5] http://nlp.stanford.edu/software/corenlp.shtml

The dependency graph [11] is a kind of syntactical structure expressed in terms of head-modifier (also called dependency) relations between pairs of words, a head and a modifier [12]. Therefore, it's easy to extract *"subject-predicate-object"* triples from the dependency graph. Furthermore, in order to extract semantic relations based on dependency graph, we should firstly generate a set of keywords which corresponds to the relationship[1] we are interested in.

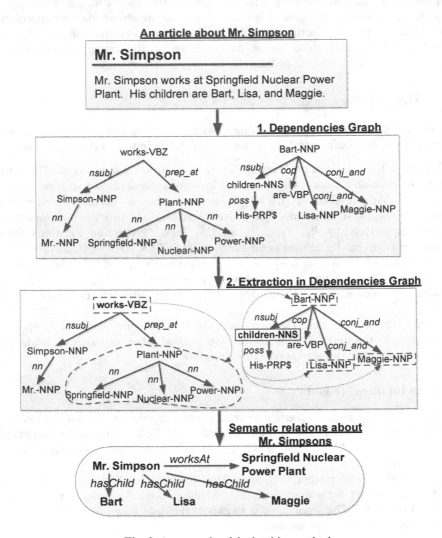

**Fig. 2.** An example of the intuitive method

---

[1] Relationship is the conceptual level of relation; that is to say, relations are instances of relationships.

For the example shown in Fig. 2, the semantic relationship "*worksAt*" is concerned with the keyword "*works*". We find "*works*" in the graph, and fetch its "*prep_at*" element as object. Then we get a semantic relation "**<Mr. Simpson>** *<worksAt>* **<Springfield Nuclear Power Plant>**". Similarly, "*hasChild*" is closely related to "*children*". However, the sentence "*His children are Bart, Lisa, and Maggie.*" is the structure of "*copula-object*", so we should extract the parent node of "*children*" as object. Thus, we get "**<Mr. Simpson>** *<hasChild>* **<Bart>**". Finally, we have to make up the "*hasChild*" relation through "*conj_and*". As a result, we get two other triples: "**<Mr. Simpson>** *<hasChild>* **<Lisa>**" and "**<Mr. Simpson>** *<hasChild>* **<Maggie>**".

The key to this extraction method is how to generate keyword set. In Section 4, we use two diffident strategies to generate keyword set and compare their results with each other.

## 3.2   Our Ontology-Based Approach

In fact, Definition 1 is not easy to follow strictly, because its condition $p_i \sim c - s - o$ is hard to realize. Thus, we came up with an ontology-based approach to our motivation. Our ontology-based method is formalized by Definition 2.

**Definition 2.** Based on the symbol settings in Definition 1, our "Ontology-Based Approach to Extracting Relations from Descriptive Text" can be defined as the function $f' : P(\mathbb{T}) \times \{\mathbb{O}\} \to P(\bigcup_{i \in \mathbb{N}^+} R_i)$. If $f(t, \mathbb{O}) = \bigcup_{i \in \mathbb{N}^+} r_i$ $(t \subseteq \mathbb{T}, r_i \subseteq R_i)$, then for any $\langle s, o \rangle \in r_i$ $(i \in \mathbb{N}^+)$, we can find $s, o \in t$, $p_i \in \mathbb{O}$, as well as some $s_0, o_0 \in \mathbb{T} \cap \mathbb{O}$ with $\langle s_0, p_i, o_0 \rangle \in \mathbb{O}$ and $\{c - s_0 - o_0 | c \in t[s_0, o_0]\} \cap \{c - s - o | c \in t[s, o]\} \neq \emptyset$.

Definition 2 is an approximate realization of Definition 1. Specifically, the condition $p_i \sim c - s - o$ in Definition 1 is replaced by $\{c - s_0 - o_0 | c \in t[s_0, o_0]\} \cap \{c - s - o | c \in t[s, o]\} \neq \emptyset$ in Definition 2. Obviously, the latter condition is much easier to be figured out than the former one.

The specific realization of Definition 2 is shown in Fig. 3. The approach that we proposed is ontology-based, and it is divided into two parts.

The first part is called *learning*. In this part, templates will be learned from the original descriptive text with the help of existing relations in the ontology. The second part is *matching & fetching*: using templates learned in the former part to match other text, and fetch information which will be reorganized into the ontology. More details about these two parts are shown in Fig. 3.

**Learning.** Learning is composed of two successive procedures. The first one is to parse the original text into parse tree. This can be figured out by the Stanford CoreNLP mentioned above.

In fact, dependency graph is likely to be more portable for learning than parse tree theoretically, as it keeps only essential information and is better organized in structure. However, dependency graph is generated based on parse tree, which means the former one has potential of deviation accumulation. Consequently, we would rather choose parse tree than dependency graph.

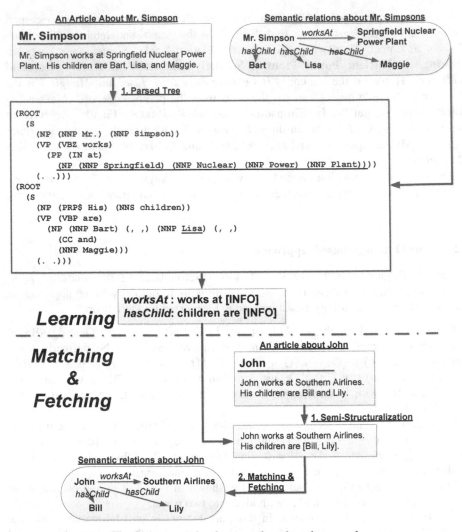

**Fig. 3.** An example of our ontology-based approach

The second one is learning templates from the parse tree, according to related triples in the ontology.

For the instance shown in Fig. 3, assuming that there are only 2 triples in the ontology: "<**Mr. Simpson**> <*worksAt*> <**Springfield Nuclear Power Plant**>" and "<**Mr. Simpson**> <*hasChild*> <**Lisa**>", we respectively find "**Springfield Nuclear Power Plant**" and "**Lisa**" in the parse trees. For the former item, it is controlled by the verb phrase "*works at*". (This can be known by tracking up the tree to find "*VP*" tag.) So we get the first template "works at [INFO]". As for the latter one, it's controlled by the verb phrase "*are*". However, the "*NP*" node before "*are*" includes a "*PRP$*" node which denotes possessive pronoun. Thus, it's necessary to append the elements after "*PRP$*" node to the template. Finally, we get the second template "children are [INFO]".

The realization of learning is shown in Algorithm 1.

**Algorithm 1.** Learn a template from parse tree, with the help of property value "item". Among the table, function "trackUpUntilTag" means tracking the tree up from current node until meeting a tag whose name is the second argument; nouns list is an enumeration of nouns; function "learnTemplateAfterTag" learns the text template after a tag whose name is the argument, and its learning procedure is same as the function "learnTemplate".

---

**Algorithm: learnTemplate**(*parseTree*, *item*)

**Input:** parse tree of the original text *parseTree*; property value *item*.
**Output:** learned template used to match relations in plain text.

1.   List *template*   :=   **new** List();
2.   **var** *node*   :=   *parseTree*.find(*item*);
3.   **var** *vp*   :=   *node*.trackUpUntilTag(*item*, "VP");
4.   **var** *np*   :=   *vp*.getFormerSilbing(*vp*);
5.   **var** *stack*   :=   **new** Stack();
6.   *stack*.push(*vp*);
7.   **while** *stack* is not empty **do**
8.        **var** *current* := *stack*.pop();
9.        **if** *current* is *item* **then break;**
10.       **elif** *current* is "adjective", "adverb", "quantity" or "determiner" **then**
11.            *template*.push(**new** Label(*current*.tag()));
12.       **elif** *current* is "clause" and *current* includes *item* **then**
13.            *template*.push(**new** Label(*current*.tag()));
14.       **elif** *current* is nouns list and *current* doesn't include *item* **then**
15.            List *nouns* = *current*.getNounList();
16.            *template*.push(**new** Label(*nouns*));
17.       **elif** *current* is nouns list **then break;**
18.       **else**
19.            **for each** *subTree* in *current*.subTrees() **do**
20.                 *stack*.push(**new** Label(*subTree*));
21. **if** it breaks from above **then**
22.       *template*.push(**new** Label([INFO]));
23.       **if** *np* includes "PRP$" **then**
24.            List *possTemplate* = *np*.learnTemplateAfterTag("PRP$");
25.            *template*.insertAt(*possTemplate*, 0);
26. **return** *template*;

---

The program above ignores adjective, adverb, quantity, determiner and irrelevant clauses, in order to be more adaptable. We believe that such process makes sense, because most modifiers make few contributions to the expression of a sentence.

**Matching.** Matching is also a pair of successive steps. The first step is to semi-structuralize the original text. This step analyzes the text, and converts it into semi-structural form, as shown in Fig. 3. The concrete algorithm for this step is exactly same as Algorithm 1, except that it doesn't care about "item".

Take what are shown in Fig. 3 as examples, "John works at Southern Airlines" is semi-structuralized to be "John works at Southern Airlines", the same as its original text, while "His children are Bill and Lily" is semi-structuralized to be "His children are [Bill, Lily]".

**Algorithm 2.** Match the template with the semi-structural text, and fetch information. Among the table, ignorable elements can be as follows: adjective, adverb, quantity, determiner and clauses.

---

**Algorithm: matchAndFetch**(*template*, *semiStructuralText*)

---

**Input:** a text *template* which corresponds to a specific relation; a sequence of labels representing *semiStructuralText*.

**Output:** all information fetched in *semiStructuralText* by *template*.

1.   List *infos*  :=  **new** List();
2.   *template*.deleteIgnorableElements();
3.   *semiStructuralText*.deleteIgnorableElements();
4.   **int**  $i := 0$,  $j := 0$,  $k$;
5.   **while** $i < semiStructuralText$.size() **do**
6.       $k := i$;
7.       **while** $j < template$.size() **do**
8.           **if** *semiStructuralText*[$k$] is NP list and *template*[$j$] is NP list **then**
9.               **if** *semiStructuralText*[$k$] and *template*[$j$] share
                            at least one same element **then**
10.                  $k := k + 1$;   $j := j + 1$;
11.              **else break;**
12.          **elif** *semiStructuralText*[$k$] is NP list **then**
13.              **if** *semiStructuralText*[$k$] includes
                          *template*[$j, j + 1, \ldots, j + u$] **then**
14.                  $k := k + 1$;   $j := j + u + 1$;
15.              **else break;**
16.          **elif** *template*[$j$] is NP list **then**
17.              **if** *template*[$j$] includes
                          *semiStructuralText* [$k, k + 1, \ldots, k + u$] **then**
18.                  $k := k + u + 1$;   $j := j + 1$;
19.              **else break;**
20.          **elif** *semiStructuralText*[$k$] == *template*[$j$] **then**
21.              $k := k + 1$;   $j := j + 1$;
22.          **else break;**
23.      **if** $j == template$.size() **then**
24.          $i := k$;
25.          *infos*.push(*semiStructuralText*[$i$]);
26.      **else**
27.          $i := i + 1$;
28.  **return** *infos*;

---

One more thing to mention, semi-structuralization must be run on all clauses respectively. This is different from that in learning part, because the latter one is run on a clause or the main clause which contains "item".

The second step is to match templates with semi-structural text and fetch the information corresponding to "[INFO]" in templates. The information will be made up with the described entity, as an instance of the relationship which would be "*worksAt*" or "*hasChild*" in Fig. 4.

For example, template "works at [INFO]" matches a semi-structural sentence "John works at Southern Airlines", and fetches "Southern Airlines". Thus, we get a semantic relation "**<John>** *<worksAt>* **<Southern Airlines>**". Similarly, template "children are [INFO]" matches "His children are [Bill, Lily]", and fetches "Bill" as well as "Lily". As a result, we get two more relations: "**<John>** *<hasChild>* **<Bill>**" and "**<John>** *<hasChild>* **<Lily>**".

The realization of matching and fetching is shown in Algorithm 2.

# 4     Experiment and Evaluation

In this section, we show how we experimented and evaluated on each method for extracting semantic relations mentioned above.

## 4.1     Experiment

Our experiments were run on the DBpedia dataset which is an open-access, high-quality and widely-used dataset derived from Wikipedia. Among the DBpedia, two sub-datasets were used in our experiments: Ontology Infobox Properties dataset and Extended Abstracts dataset. The former one provided us with an existing ontology, while the latter one which contains abstracts in Wikipedia, offered us a large amount of descriptive texts.

Experiments were run in relationships; that is to say each experiment's results must be about one and only one specific relationship. Thus, before running our experiments, we chose 20 properties randomly. After that, we randomly chose no more than 300 subjects for each relationship respectively. As a result, a dataset for our experiments was created, which contains 20 relationships with no more than 300 subjects for each.

**Experiment Setup for the Intuitive Method.** As mentioned in Subsection 3.1, how to generate the keyword set is critical and makes a great difference on its performance. In our experiments, we came up with two strategies to get the keyword set.

The first one was taking synonyms of relationship's name and their different forms as the keyword set. For example, the relationship "*hasChild*" is synonymous to "child", "baby", "kid", and etc. These synonyms have other forms, such as "children", "babies", "kids", and etc. We took words above as the keyword set of "*hasChild*". Incidentally, synonymous were found by a large English lexical corpus, named WordNet [13].

The second one was extended from the first one. The extension came from all relation templates extracted by the ontology-based approach.

**Experiment Setup for the Ontology-Based Approach.** As is an ontology-based method, a dataset has to be divided into training set and test set. As described in Subsection 3.2, the approach firstly extracted relation templates from the training set, and then matched the test set with the extracted templates, so as to fetch objects of the relationship being experimented with.

In our experiments, we divided the data set into 5 subsets, and took one subset as the training set and the rest as the test set in turns. Such practice is the well-known 5-fold cross-validation.

## 4.2   How to Evaluate

We evaluated our methods with two important measures: recall & precision.

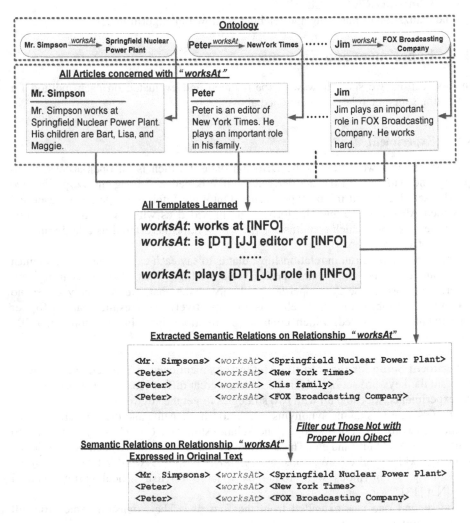

**Fig. 4.** An example of how to approximately fetch "Origin" set

$$recall = |Extracted \cap Origin|/|Origin| \; ; \tag{1}$$

$$precision = |Extracted \cap Origin|/|Extracted| \; . \tag{2}$$

Here, "Extracted" represents those triples extracted from the test set, while "Origin" is composed of the triples that are exactly what the test set expresses.

For the experiment of the ontology-based approach, recall, as well as precision, was figured out on each fold, and the average was taken as the final value.

It seems to be easy to figure out these measures. However, it is not, because we do not have the data of "Origin". "Origin" does not equal to the ontology. It is apparent that some triples are in "Origin" but not in the ontology, while some are in the ontology but not in "Origin". Such dilemma made it difficult to do statistics on the results.

Fortunately, we came up with an idea to deal with it. The kernel idea is to make use of the particularity of the situation where relationship's object type is proper noun. Objects are believed to be in "Origin" with high probability, when they are not only proper nouns, but also fetched by relation templates.

With the assumption above, a method to approximately fetch "Origin" set forms, as Fig. 4 shows: Firstly, templates were extracted from all descriptive text concerned with "*worksAt*". Then, all templates were used to match back with what they had been extracted from. Semantic relations with predicate "*worksAt*" were fetched and those not with proper noun object would be filtered out. As a result, the left semantic relations above were considered to be elements in the "Origin" set approximately.

## 4.3    Results

Three teams of experiments were carried out on each relationship respectively. These teams were ontology-based approach, the intuitive method with only synonyms as keyword set, and the intuitive method with all relation template keywords added to the former one as keyword set. More detail can be found in Subsection 4.1.

In order to make it more clear, here is an example of "all relation template keywords": if the relationship is "*worksAt*" as in Fig. 4, its relations template keywords are templates' meaningful words, and here they are "works", "editor", "plays" and "role".

As mentioned at the beginning of this Section, 20 relationships had been chosen randomly for our experiments. However, among these relationships, there were 5 not owning proper noun object type. Thus, we had to give up evaluations on these 5 relationships. Finally, evaluations were conducted on each team among the rest 15 relationships.

**Recall.** Recall on each approach and each relationship is shown in Fig. 5. We found that our ontology-based method performs significantly better than the intuitive method for most relationships. We believe this is because the former method is much less dependent on syntactical structure than the latter one. Besides, for the intuitive method, taking synonyms and all relation template keywords as keyword set performs much better than only using synonyms. It proves that most semantic relations are not expressed with their relationship's names in plain text, which makes it difficult to generate proper keyword set for this method.

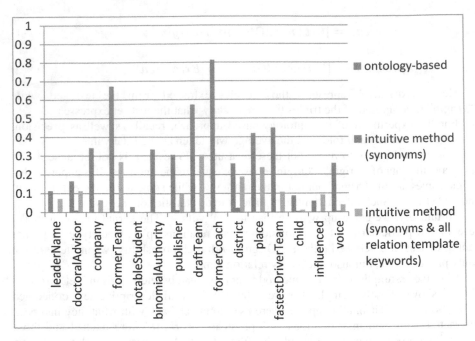

**Fig. 5.** Recall on each approach and each relationship

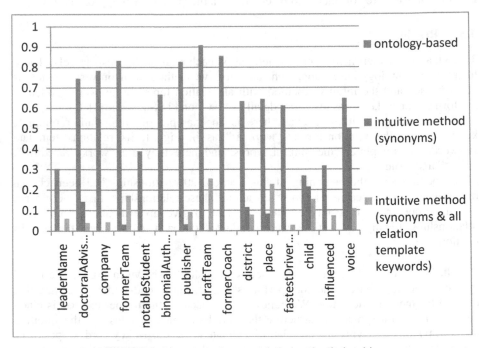

**Fig. 6.** Precision on each approach and each relationship

**Precision.** Precision on each approach and each relationship is shown in Fig. 6. We found that our ontology-based method also performs much better than the intuitive method for all tested relationships. We believe this is because templates used by the former method are much stricter than keywords used by the latter one. Besides, for the intuitive method, taking synonyms and all relation template keywords as keyword set doesn't absolutely perform better than only using synonyms. It can be presumed that extending keyword set can lead to more incorrect information extracted.

At last, the average performance on each approach is shown in Table 1.

**Table 1.** The average performance on each approach

| Approach | Recall | Precision |
|---|---|---|
| ontology-based | 0.325 | 0.629 |
| intuitive method (synonyms) | 0.004 | 0.086 |
| intuitive method (synonyms & all relation template keywords) | 0.107 | 0.089 |

## 5    Conclusions and Future Work

In this paper, we put forward an ontology-based approach to extracting semantic relations from descriptive text. The approach learns the relation templates with the help of ontologies, and matches plain texts with the learned templates in order to extract semantic relations from them.

We compare this approach with the intuitive method by using recall and precision. It is demonstrated that our ontology-based approach performs better than the intuitive method as shown in Fig. 5-6 and Table 1. The results may be due to:

1. The intuitive method's deviation accumulation on dependency graphs.
2. The intuitive method not good at dealing with complex syntactical structures.

However, better performance here does not mean that our ontology-based approach wins the intuitive method in all conditions. The intuitive method can be a better choice when the ontology is extremely rare.

This work is just the first step towards annotating semantic relations from plain text. In future work, several measures will be researched to improve the performance of the ontology-based approach. These measures can be:

1. Take relationship's domain and range into account, and restrict the class of values when being matched with relation templates.
2. Introduce fitting times into relation templates. Only those with high fitting times will be chosen as usable templates. Here, the value of fitting times indicates how many times a relation template is fit in training set.
3. Combine the intuitive method with the ontology-based one. Take semantic relations extracted by the intuitive method as candidates. Then, provide users with these candidates to choose. Those chosen relations will be taken as positive examples for learning relation templates.

We believe the first two measures can help improve precision, while the last one may make a difference in recall.

**Acknowledgements.** This work is supported in part by the National Natural Science Foundation of China (Nos. 61223003 and 61003018), in part by the National Social Science Foundation of China (No. 11AZD121), and in part by the Natural Science Foundation of Jiangsu Province (No. BK2011189).

# References

1. Bizer, C., Heath, T., Berners-Lee, T.: Linked Data – The Story So Far. International Journal on Semantic Web and Information Systems 5(3), 1–22 (2009)
2. Auer, S., Bizer, C., Kobilarov, G., Lehmann, J., Cyganiak, R., Ives, Z.G.: DBpedia: A Nucleus for a Web of Open Data. In: Aberer, K., Choi, K.-S., Noy, N., Allemang, D., Lee, K.-I., Nixon, L.J.B., Golbeck, J., Mika, P., Maynard, D., Mizoguchi, R., Schreiber, G., Cudré-Mauroux, P. (eds.) ASWC 2007 and ISWC 2007. LNCS, vol. 4825, pp. 722–735. Springer, Heidelberg (2007)
3. Buitelaar, P., Cimiano, P., Magnini, B.: Ontology Learning from Text: An Overview. ACM Computer Surveys. In: Breuker, J., et al. (eds.) Ontology Learning from Text: Methods, Evaluation and Applications. Frontiers in Artificial Intelligence and Applications, vol. 123, pp. 3–12. IOS Press, Amsterdam (2005)
4. Vela, M., Declerck, T.: A Methodology for Ontology Learning: Deriving Ontology Schema Components from Unstructured Text. In: Handschuh, S., et al. (eds.) Workshop on Semantic Authoring, Annotation and Knowledge Markup 2007. SEUR-WS, vol. 289. CEUR-WS.org, Whistler (2007)
5. Cimiano, P., Hotho, A., Staab, S.: Learning Concept Hierarchies from Text Corpora using Formal Concept Analysis. Journal of Artificial Intelligence Research 24(1), 305–339 (2005)
6. Wong, W., Liu, W., Bennamoun, M.: Ontology Learning from Text: A Look Back and into the Future. ACM Computing Surveys 44(4), 20 (2012)
7. McDowell, L.K., Cafarella, M.: Ontology-Driven Information Extraction with OntoSyphon. In: Cruz, I., Decker, S., Allemang, D., Preist, C., Schwabe, D., Mika, P., Uschold, M., Aroyo, L.M. (eds.) ISWC 2006. LNCS, vol. 4273, pp. 428–444. Springer, Heidelberg (2006)
8. Khelif, K., Dieng-Kuntz, R., Barbry, P.: An Ontology-based Approach to Support Text Mining and Information Retrieval in the Biological Domain. Journal of Universal Computer Science 13(12), 1881–1907 (2007)
9. Agichtein, E., Gravano, L.: Snowball: Extracting Relations from Large Plain-Text Collections. In: 5th ACM Conference on Digital Libraries, pp. 85–94. ACM Press, New York (2000)
10. Fundel, K., Kuffner, R., Zimmer, R.: RelEx – Relation extraction using dependency parse trees. Bioinformatics 23(3), 365–371 (2007)
11. Barbero, C., Lombardo, V.: Dependency graphs in natural language processing. In: Gori, M., Soda, G. (eds.) AI*IA 1995. LNCS, vol. 992, pp. 115–126. Springer, Heidelberg (1995)
12. Lesmo, L., Lombardo, V.: The assignment of grammatical relations in natural language processing. In: 14th Conference on Computational Linguistics. Project Notes with Demonstrations, vol. 4, pp. 1090–1094. Association for Computational Linguistics, Stroudsburg (1992)
13. Miller, G.A.: WordNet: a lexical database for Engilish. Communications of the ACM 38(11), 39–41 (1995)

# Linking Biomedical Data
# for Disease–SNP Relation Discovery

Na Hong, Qing Qian, An Fang, Sizhu Wu, and Junhui Wang

Institute of Medical Information
Chinese Academy of Medical Sciences, Beijing, China
{hong.na,qian.qing,fang.an,wu.sizhu,wang.junhui}@imicams.ac.cn

**Abstract.** Traditional relation discovery is always conducted through either text mining or database analysis. However, in the real world, knowledge exists in different formats and can be expressed in a variety of forms. Discovering relations between diseases and single-nucleotide polymorphisms (SNPs) is challenging because of difficulties in unstructured data processing or distributed heterogeneous data integration. With the development of Sematic Web theory and technology, it provides feasibility to reconstruct the traditional data integration process in a sematic manner in the biomedical big data era. Our study aims to discover disease–SNP relation in integrated linked data to facilitate scientific research analyses and reduce biological experiment costs. We focus on investigating the capability of linked data techniques in integrating and mining relationships between diseases, genes, and SNPs. To demonstrate the effectiveness of our proposed method, we conducted a case study in Alzheimer's disease-SNPs discovery by integrating 10 datasets.

**Keywords:** Semantic Web, Linked Data, SNPs, Relation Discovery.

## 1    Introduction

With the development of biomedicine and humans' continuous exploration in it, a great amount of information and knowledge have been created. Increasing data in different fields are always stored in special databases respectively. In most cases, however, they are heterogeneous and with different formats. These data can be available through various organizations and exist in a variety of formats, sizes and types. Nevertheless, such data are essentially implicated and actually seperated in many cases. Discovering the underlying relationships among such data can facilitate research analysis and reduce the cost of biological experiments.

Many research progresses have already been made on knowledge relationship discovery. Most current methods of knowledge relation discovery are based on text mining and database integration. Text mining algorithms and tools, including information extraction, semantic tagging, and relation extraction, promote knowledge relationship discovery. In addition, many database integration methods, such as federated search and data schema mapping, are used to do this work.

G. Qi et al. (Eds.): CSWS 2013, CCIS 406, pp. 37–49, 2013.
© Springer-Verlag Berlin Heidelberg 2013

In recent years, Semantic Web technology has introduced a new approach for knowledge discovery research. Integrating large-scale biomedical data through linked data can ensure resource integration and sharing, allow scientists to conduct thorough analysis and exploration, create new knowledge, and discover underlying knowledge.

We integrated a large number of disease, gene, and single-nucleotide polymorphism (SNP)-related data sources into linked data and designed a system which could automatically discover the underlying relation between a specific disease–SNP pair. The discovered paths are then ordered by using a multi-dimension ranking algorithm.

## 2     Background

In recent years, linked data has been considered a practicable method to realize Semantic Web. Linked data is also proposed as an effective way to solve semantic problems in this era of biomedical big data.

### 2.1     The Need to Discover Disease–SNP Relation

Recent findings indicate that SNP can lead to susceptibility to and onset of diseases through their effects on gene expression at the post-transcriptional level. Although a large number of candidate genes currently exist, candidate SNPs and disease are related in essential. However, detecting their relation is difficult, especially for complex diseases like cardiovascular disease which always have many related candidate genes and SNP. In most cases, studies that focused on an isolated specific gene or SNP cannot explain the relation among them. In addition, experiment on each candidate genes and SNPs is costly. If numerous candidate genes exist, identifying the gene type of all candidate SNPs within the candidate genes is impossible. Otherwise, discovering disease candidate SNPs is meaningful to pharmacogenomics research [1]; it is helpful for screening drug targets and save the time and cost of drug dicovery. So, we attempt to identify disease-related SNPs through automatic relation discovery technologies and then provide the candidate SNP sample for biological experiment.The Genome-wide Association Studies (GWAS) of the National Institute of Health (NIH) is also conducting research on SNP and disease relation analysis [2].

### 2.2     Relation Discovery Study Based on Linked Data

Linked data are valuable in drug discovery, life sciences, and biomedical research [3,4,5]. A number of studies and applications verified its feasibility and provided a significant reference for our study. Chem2Bio2RDF [6], BioNav [7], Pathfinder [8] and OpenPHACTS [9] are some of the typical achievements in this field.

Chem2Bio2RDF is a linked data integrated by the chemical, biological and medicine data sets. Data are stored in an RDF triples, and applied to find the connecting paths between two entities or concepts. Usually, a pair of entities or concepts belonging different environments could be connected through different link paths, so they will carry different interpretations. They tried to find out all the "chemical compounds - disease" association, for instance, they found 81077 different chemical compounds are associated with alzheimer disease, and 410 of which connected through the bridge builded by the specific gene.

BioNav is a framework and system which detects latent semantic links from linked data cloud, and it was designed based on ontology technology, which support BioNav effectively detecting potential, novel relationship between drug and disease. By exploring the large scale of linked data cloud with ontology and the ranking techniques, BioNav could return the top n links for detail analysis. The main BioNav ranking technology is based on link analysis and ranking matrix. Test proof that BioNav can find most of the effective relationship.

Pathfinder is a complicated relation discovery system towards web triples, and it could find the complicated relationship between two entities seemingly unrelated. The heuristic method is used to measure the relationship and remove loop interference. After the experiment for the discovery of simple entity relationships such as person, organization and location which states it can successfully predict potentially 6 kinds of relations.

OpenPHACTS project follow the idea of building an open, integrated, large, semantic repository of drug discovery. It has builded an extensive, precompetitive Open Pharmacological Space (OPS) of linked data for drug discovery.OPS has created many effective relationships between drug data, and made some successful applications in the filed of drug discovery and pharmaceuticals industry.

## 2.3    System Requirement

The representation format, semantic ability, storage capacity, publish pattern, and query efficiency of linked data are the foundations of our research. Therefore, we list the core requirements to support subsequent studies and experiments.

To unified store data with diffirent format, data should be converted into a unified format first, and in most data conversion cases, direct conversion is more efficient than conversion by using a third-party tool. Then, to ensure that large volumes of data can be stored, RDF storage capacity and query speed must be guaranteed. Furthermore, to discover the underlying relation, the superficial relations must be established. Therefore, connecting instances through mapping stratege is a vital issue. The following steps are focused on fixing these preparing problems before relation discovery research:

(1)  Thoroughly analyze the content and structure of the original data, and then directly extract entities and relations from fields in original database to prepare drug discovery relevant data.
(2)  Assign a unique HTTP URI to all objects; this is one basic requirement of linked data
(3)  Use link URI to keep data linked as much as possible
(4)  Try to avoid data redundancy
(5)  Create mapping rules before mapping
(6)  Ensure large-scale storage can contain up to more than 1 billion triples

(7) Provide a convenient query interface to ensure that data are loaded within a short time, as well achieve an acceptable query response time and query stability

(8) Ensure that the Resource Description Framework (RDF) repository has enough reasoning ability to support the semantic inference

## 2.4    Semantic Technology Analysis

Our research goal is to link data in biomedical datasets and discover any knowledge relevant to diseases and SNP in the Semantic Web environment. In the Semantic Web environment, knowledge expression, storage, organization, and consumption are different from those in the World Wide Web. After our analyzing on semantic technology, such as RDF, OWL, RDFizer, SPARQL, and N-triple storage, we considerd the following issues should be especially studied:

(1)The key difficulty in integrating data in biomedical datasets is that data usually be organized in different standards and forms. The recommended storage formats of linked data for W3C are RDF/XML, N-triple, Turtle, and JSON. Therefore, identifying the strategies and tools for converting non-RDF data into RDF is the primary problem. Conversion tools differ among different data formats, such as Any23 (Anything To Triples), D2R, and Drupal. Actually, we opted to convert data through a conversion program we designed ourselves based on data analysis.

(2)Determining how to solve the problems of RDF data storage and access is essential in a linked data application. Existing studies that used Triplestore, RDF Store, and Semantic Repository, all attempted to improve storage capacity and query efficiency. We found that AllegroGraph, OpenLink Virtuoso, and BigOWLIM have outstanding storage capacity, and Jena TDB and Virtuoso have high query efficiency with a small number of triples. However, TDB has inadequate storage capacity. After our analysis and test, we regarded Virtuoso possessing both excellent storage capacity and speed.

(3)Linked data often contain extensive resources and can be linked to hundreds of thousands of instances and attributes. Therefore, visualization tools are neccesary to help browse linked data. RDF visualization can allow a user to understand the complex internal structure between entities easily. In addition, it can depict the details of any discovered relations. Therefore, we analyzed a number of RDF visualization tools, including RelFinder, RDF Gravity, Isaviz, graphviz, Gruff, and OntoSphere. RelFinder performs best in terms of interactive relation discovery and was therefore selected as the relation display platform for our system.

## 3    Methods

In this article, we call our system as DSRedis (Disease- SNP Relation Discovery System based on Linked Data). In the subsequent sections we will describe the framework and the main parts of the DSRedis.

### 3.1   DSRedis Framework

The DSRedis was built upon an automatic discovery framework which was created through linked data integration, storage, querying, publishing, and inference. The core functions were designed to perform relation discovery. The framework involves the following process: the user inputs a pair of entities that are sented into the linked data repository for query. An automatic search in the linked data repository performed, and then DSRedis can recognize links between two entities. Links are then ordered according to ranking algorithm, multiple parameters are calculated to measure the importance of the link path. DSRedis then presents hypotheses, a direct explanation, or an indirect inference about relations based on the link paths. Also, DSRedis can present discovered links visually and allow the user to explore the system interactively. (Fig. 1)

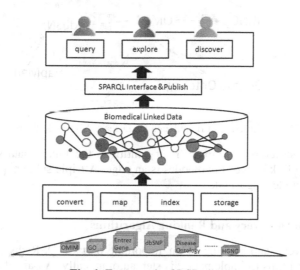

**Fig. 1.** Framework of DSRedis

### 3.2   Biomedical Linked Data Integration

To discover the relation between a specific disease and SNP by using our method, linked data must be constructed based on disease, gene, SNP, and other related information. The basic goal of linked data population is to associate these different data, annotate the relationships with a detailed label between isolated entities, and then publish the data for knowledge sharing and discovering.

The system primarily involves building concepts and entities within biomedical linked data from a number of well-established biomedical ontologies and databases. To find disease–SNP relations, we collected data related to genetics and disease as much as possible. First, we tried to creat an underlying knowledge framework by Gene Ontology and Disease Ontology. Then, data from OMIM, HapMap, HGNC, dbSNP, diseasome, miRdSNP, PubMed, and EntrezGene were integrated into this framework. All data are derived from different sources and stored in different formats, such as n3, Excel, XML, BCP, FASTA, and SQL database.

Next step, the data were converted into unified n3 format. Thousands of entities with mappings were connected as much as possible. Three basic methods were used during mapping process, they are ID matching, name matching, and matching according to the document provided by the original dataset. By using the above methods, we populated the biomedical linked data to discover the disease–SNP relation. (Fig. 2)

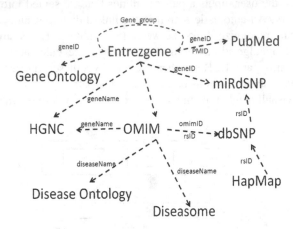

**Fig. 2.** Structure of linked data for Disease-SNP link discover

Linked data structure is a graph. Every entity in the linked data we created is a node, and every link between two nodes is an edge. A triple subject–predicate–object corresponds to node–edge–node.

### 3.3    Relation Discovery and Ranking Algorithms

Relation discovery and ranking is an important part of DSRedis. We selected a number of algorithms to implement this step automatically. A pair of entities may be linked directly or indirectly. In most cases, links are indirect and are composed of more than one-step edge. We focused especially on the indirect links on the assumption that the underlying relation always be found through indirect links. Therefore, we provided detailed path information of every edge to distinguish which links are reasonable and significant.

How all underlying relations between two entities in linked data are found is another problem. In this case, a search algorithm is used to find all edges between two nodes in Triple Store. Here, we adopt a series of SPARQL queries to achieve graph traversal.

First, all paths are found and then ranked in different angles to filter the irrelevant paths and identify the relevant ones. We investigate related ranking strategy, and we adopt four different parameters, namely, path length, edge weight, rarity degree, and degree of popularity [8].

(1) Path length. Short path is ranked higher than long path. Assume that A is a path, Rank (A) = 1 / length (A).

(2) Edge weight. Each edge is assigned a weight. Assume that q1 to qn are represent of the weights of each edge in a path, the path A ranked by:

Rank (A) = (1/length(A))(q1+q2+q3+...+qn).

(3) Rare degree. This parameter needs to calculate the rare degree of all nodes and edges composed a path. Assume that each node or edge is an object; all objects need to be clustered at first.

Rare(object)  = M-N/M

M is the total number of all objects,  N is the total number of objects in its cluster, and the path A ranked by:

Rank (A) =(1/length(A))(Rare(object(1))+Rare(object(2))+...+ Rare(object(n)))

object(1) to object(n) represent all nodes and edges composed the path A.

(4) Degree of popularity. This parameter needs to calculate the number of input edges and output edges for each node.

Popular (node) =num (input edge) + num (output edge)

After every node's popular degree has been calculated, we can give the rank value for path A:

Rank(A)=(1/length(A))(Popular(node(1))+Popular(node(2))+...+Popular(node(n)))

node(1) to node(n) represent all nodes exist in path A.

# 4     Experiment and Results

## 4.1     Linked Data Statistics

In DSRedis, biomedical datasets were converted into RDF and then integrated as linked data. A total of 22,089,203 triples were stored in Openlink Virtuoso. For SPARQL query and storage scale, Virtuoso has good performance as triplestore to support DSRedis. Detailed information about data in our system is shown as Table 1. Furthermore, DSRedis created a set of mapping relations between these data sets. The system created 12,528,322 mappings in the form of triples and then stored in Virtuoso. Semantic labels were used to represent those relationships. Parts of the created mapping relations are shown in Table 2.

**Table 1.** Source and size of linked data in DSRedis

| Dataset | Source | Named Graph | Triples |
| --- | --- | --- | --- |
| DiseaseOntology | Nugene Project | http://localhost:8890/diseaseontology | 184348 |
| HGNC | NHGRI | http://localhost:8890/hgnc | 1702786 |
| Symptom | Gemina Project | http://localhost:8890/symptom | 3345 |
| GeneOntology | GO Consortium | http://localhost:8890/geneontology | 320165 |
| Diseasome | Goh K, Cusick M, Valle D, etc. | http://localhost:8890/diseasome | 72445 |
| OMIM | NCBI | http://localhost:8890/omim | 17251 |
| HAPMAP | NCBI | http://localhost:8890/hapmap | 23466 |
| dbSNP | NCBI | http://localhost:8890/dbsnp | 428851 |
| miRdSNP | Center for Computational Research SUNY at Buffalo | http://localhost:8890/mirdsnp | 1896244 |
| EntrezGene | NCBI | http://localhost:8890/entrezgene | 4802367 |

**Table 2.** Mapping rules created and labels used in DSRedis

| Source | Target | Mapping rule | Semantic_label |
|---|---|---|---|
| OMIM | HGNC | Gene ID matching | Same_genename |
| OMIM | DiseaseOntology | Disease name matching, Gene ID matching | Gene_disease |
| OMIM | Diseasome | Disease name matching | Same_diseasename |
| OMIM | dbSNP | Gene ID matching, published document of NCBI | Allelic_number |
| OMIM | EntrezeGene | Gene ID matching, published document of NCBI | Phenotype/gene |
| GeneOntology | EntrezeGene | Gene ID matching | Same_geneID |
| EntrezGene | EntrezeGene | Gene ID matching, published document of NCBI | Gene_rel |
| EntrezGene | PubMed | Gene ID matching, published document of NCBI | Gene_ref_pubmed |
| DiseaseSome | EntrezeGene | Gene name matching | Same_genename |

## 4.2    DSRedis Implementation

Alzheimer's disease is used as an example. When we input "Alzheimer" in DSRedis, the system will provide a number of related terms for further selection. Assuming that we choose the term "Alzheimer's disease" and input another term, which is any one of the SNP numbers such as "rs63750643," DSRedis will return an ordered path by ranking a series of path search results, ranking can be ordered in five dimentions, Besides four paremeters in 3.3, we also give a TotalScore based on weighted average methods of the four paremeters, as shown in Fig.3. For every node or edge, we just displayed the last filed of URI as the abbreviation of them. In the directed path, edges with green color represent right arrows, and edges with red color represent left arrows.

| | Path | TotalScore | Length | Weight | RareDegree | PopularI |
|---|---|---|---|---|---|---|
| 1 | 1403 associatedGeneAPOE same_genename348 type_of_geneprotein-coding type_of_gene351 gene104760 15324rs63750643 | 56 | 6 | 65 | 62 | 80 |
| 2 | 1403 diseaseSubtypeOf74 associatedGeneAPP same_genename351 gene104760 15324rs63750643 | 51.8 | 5 | 78 | 73.5 | 35.1 |
| 3 | 1403 diseases_gene107741 typeomim_gene type104760 15324rs63750643 | 50.3 | 4 | 60 | 67.3 | 46.5 |
| 4 | 1403 diseaseSubtypeOf74 associatedGeneAPP same_genename351 same_geneidHGNC%3A620 same_genename104760 15324rs63750643 | 49.2 | 6 | 80 | 69.6 | 29 |
| 5 | 1403 diseaseSubtypeOf74 same_diseasename98 gene104760 gene351 gene104760 15324rs63750643 | 47 | 6 | 80 | 64.8 | 28.1 |
| 6 | 1403 diseaseSubtypeOf74 same_diseasename98 gene104760 15324rs63750643 | 45.4 | 4 | 75 | 80 | 2.9 |
| 7 | 1403 diseaseSubtypeOf74 same_diseasename98 gene107741 typeomim_gene type104760 15324rs63750643 | 44 | 6 | 60 | 62 | 32.1 |
| 8 | 1403 diseaseSubtypeOf74 same_diseasename98 gene600759 typeomim_gene type104760 15324rs63750643 | 43.9 | 6 | 60 | 62 | 31.8 |
| 9 | 1403 diseaseSubtypeOf74 associatedGeneAPP labelAPP Symbol351 gene104760 15324rs63750643 | 43.1 | 6 | 60 | 62 | 28.3 |

**Fig. 3.** Rank of path search results

To further explore the relation between two entities, interactive visualization tools need to be used. RelFinder[1] was adopted due to its high interactive visualization performance. The above path list can also be visualized by RelFinder in DSRedis.

The application of RelFinder needs only a little configuration; we realized the visualization of path through following configuration.

(1) Running environment Settings. RelFinder is a Flex based visual tools, so we need install Flex and SpringGraph. Flex enables the basic interaction function between user and interface, and SpringGraph enables the network diagram and its high complexity interactive features along with the seamless integration. Therefore, Flash Player plugin must be installed in the browser when using RelFinder.

(2) SPARQL endpoint configuration. RelFinder can be used in different RDF database through the configuration of SPARQL endpoint. SPARQL querys generate relatively independent on client and server. The server uses a series of PHP script to complete SPARQL querys submission and return results between client and SPARQL endpoint. Therefore, the only operation of using the RelFinder is to modify the config XML file. In our experiment system, we set the SPARQL endpoint URI as http://localhost:8890. The config XML file is as follows:

```
<endpoint>
          <name>Linked Biomedical Ddata</name>
          <abbreviation>LBD</abbreviation>
     <description>Linked Data version of   Biomedical.</description>
          <endpointURI>http://localhost:8890</endpointURI>
          <dontAppendSPARQL>false</dontAppendSPARQL>
     <defaultGraphURI>http://localhost:8890</defaultGraphURI>
          <isVirtuoso>false</isVirtuoso>
          <useProxy>false</useProxy>
          <method>POST</method>
     .......
          <maxRelationLength>5</maxRelationLength>
</endpoint>
```

For example, input "Alzheimer disease - 2, 104310" and "rs63749810", RelFinder can help us explore the detail path clearly based on DSRedis, as seen in Fig 4. Every node in the graph with a unique URI, However we just use the label of each subject to represent the node so as to show the relation clearly, so node labels sometimes look the same, such as APP, but essentially they are two different URIs and belonging to two different datasources( eg.HGNC and Diseasome).

---

[1]  RelFinder, Visual Data Web http://www.visualdataweb.org/relfinder.php

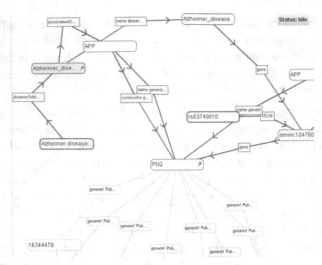

**Fig. 4.** Relation discovered between "Alzheimer disease - 2, 104310" and "rs63749810"

In all discovered paths, we select two of them to explain the biological meaning, with the triples structures detailed below:

（1）<Alzheimer disease-2, 104310> <disease_subtype> < Alzheimer Disease>
< Alzheimer Disease > <associated_gene> < APP>
<APP> <synonyms_gene> < PN2>
<APP> <same_genename> < PN2>
<Omim:104760> <gene> < PN2 >
< entrenzgene:351> < allelic_number:15218> < rs63749810>
< entrenzgene:351> <symbol> < APP>
< entrenzgene:351> <synonym> < PN2>
<Omim:104760> <same_gene> < entrenzgene:351>
（2）< Alzheimer disease-2, 104310> <associated_gene> < APOE>
<APOE> <same_gene> < LPG>
< LPG > <pubmed_id> < 16344478>
< PN2 > <pubmed_id> < 16344478>
<Omim:104760> <gene> < PN2 >
< entrenzgene:351> < allelic_number:15218> < rs63749810>
< entrenzgene:351> <symbol> < APP>
< entrenzgene:351> <synonym> < PN2>
<Omim:104760> <same_gene> < entrenzgene:351>

In the first path, we can see that a broad disease concept of "Alzheimer disease - 2, 104310" is Alzheimer disease, the gene to which Alzheimer disease related is APP, the approximate genes of APP is PN2, PN2 gene changed in 15218 loci, and the polymorphism marker is "rs63749810". In the second path, "Alzheimer disease - 2, 104310" related gene is APOE, Gene APOE and gene LPG are same, LPG and PN2 co-occur in the same PubMed article   and the PMID is16344478, PN2 gene changed in 15218 loci, and the polymorphism marker is "rs63749810". Thus, we assume that "Alzheimer disease - 2, 104310" and "rs63749810" have a certain underlying relationship.

The discovered relationship is supported by a variety of recent research which indicates that rs63749810 is implicated in Alzheimer's disease. ClinVar[2] data indicates rs63749810 is related of Cerebral amyloid angiopathy, and the chromosome position is 27,264,165. Cerebral amyloid angiopathy has been recognized as one of the morphologic hallmarks of Alzheimer's disease, but it is also often found in the brains of elderly patients who are neurologically healthy[10,11]. Cerebral amyloid angiopathy may lead to dementia, intracranial hemorrhage, or transient neurologic events. Besides, the relationship between Cerebral amyloid angiopathy and Alzheimer's disease has already been confirmed by other literature [12,13,14].

The above path discovering process can also be used in other pair of entities. That is to say, our method can be used to discover the relations between any other entities and is not limited to the relation between a disease and SNP. The discovery application is determined by data domain.

### 4.3   Discussion

The experiment clearly proved that an underlying relation exists between entities. Therefore, we consider DSRedis a reasonable way to discover underlying relations between disease and SNP based on linked data. However, the following problems still need to be discussed:

(1) The mapping effect needs to be improved. In DSRedis, the ID and name matching strategies can create limited mappings only. More effective methods should be used, such as an algorithm that considers the concept background information, context information, and makes an inference from exist relations, to create more effective semantic relations between entities or concepts, and then support effective knowledge discovery.

(2) RDF query efficiency requires further optimization, so as to support more large scale, reasonable SPARQL queries and path finding.

(3) The system needs to integrate more data sets and focus on more professional data sets. In our future work, we will analyze disease–SNP relation more deeply and take other SNP-related databases into consideration.

## 5   Conclusion

Semantic Web technology provides new opportunities as well as introduces new ideas for biomedical knowledge discovery. In recent years, knowledge expression, storage, organization, and utilization have undergone significant changes as a result of new and updated semantic technology. Linked data is the most feasible method in

---

[2] ClinVar. http://www.ncbi.nlm.nih.gov/clinvar/

Semantic Web. Following the principle of linked data, biomedical objects were further organized through a RDF graph structure which links entities or concepts to identify the relationship between heterogeneous databases and implicit knowledge. According to our investigation of genetic medical specialties and literature, our study results conform to the current disease–SNP development status and provide reasonable explanations. We therefore conclude that linked data is a new way to integrate biomedical knowledge.

**Acknowledgments.** This work is funded by the National social science fund project (NSSF), China. Project Number is 11CTQ016. We also wish to thank Chem2Bio2RDF project (University of Indiana) and Linked Life Data project (Ontotext and LarKC), and in our experiment, part of datasets were converted referring to these project.

# References

1. ChengHong, Y., YuHuei, C., LiYeh, C., HsuehWei, C.: Drug-SNPing: an integrated drug-based, protein interaction-based tag SNP-based pharmacogenomics platform for SNP genotyping. Bioinformatics 29, 758–764 (2013)
2. Wataru, S., Yuko, N., Ikuko, M., Yushi, H., Chiyomi, I., Michiaki, K., Takahisa, K., Tatsuhiko, T., Masahiko, W., Atsushi, T., Hiroyuki, T., Kenji, N., Kazuko, H., Fumiya, O., Takeo, Y., Hideshi, K., Saburo, S., Mitsutoshi, Y., Nobutaka, H., Miho, M.: Genome-wide association study identifies common variants at four loci as genetic risk factors for Parkinson's disease. Nature Genetics 41, 1303–1307 (2009)
3. Slater, T., Bouton, C., Huang, E.S.: Beyond data integration. Drug Discovery 13, 584–589 (2008)
4. Neumann, E.K.: A life science semantic web: are we there yet? Scienc Today 283, 22–25 (2005)
5. Neumann, E.K., Miller, E., Wilbanks, J.: What the semantic web could do for the life sciences. Drug Discovery Today 2, 228–234 (2006)
6. Xiao, D., Ying, D., Huijun, W., Bin, C., David, J.: Chem2Bio2RDF Dashboard: Ranking Semantic Associations in Systems Chemical Biology Space. In: FWCS(The Future of the Web for Collaborative Science), Raleigh, USA (April 26, 2010)
7. Vidal, M.-E., Raschid, L., Márquez, N., Rivera, J.C., Ruckhaus, E.: BioNav: An Ontology-Based Framework to Discover Semantic Links in the Cloud of Linked Data. In: Aroyo, L., Antoniou, G., Hyvönen, E., ten Teije, A., Stuckenschmidt, H., Cabral, L., Tudorache, T. (eds.) ESWC 2010, Part II. LNCS, vol. 6089, pp. 441–445. Springer, Heidelberg (2010)
8. Benjamin, B., Ranjana, S., Ritika, S., Phani, T.: Pathfinder: Complex Relation Discovery and Ontological Management Software for Generic Ontologies or Web Based Triples, http://obiwan.cs.ndsu.nodak.edu/~rsharma/AIProject.pdf
9. Harland, L.: Open PHACTS: A semantic knowledge infrastructure for public and commercial drug discovery research. In: ten Teije, A., Völker, J., Handschuh, S., Stuckenschmidt, H., d'Acquin, M., Nikolov, A., Aussenac-Gilles, N., Hernandez, N. (eds.) EKAW 2012. LNCS, vol. 7603, pp. 1–7. Springer, Heidelberg (2012)
10. Chung, Y.A., Hyun, O.J., Kim, J.Y., Kim, K.J., Ahn, K.: Hypoperfusion and Ischemia in Cerebral Amyloid Angiopathy Documented by 99mTc-ECD Brain Perfusion SPECT. J. Nucl. Med. 50, 1969–1974 (2009)

11. Weller, R.O., Preston, S.D., Subash, M., Carare, R.O.: Cerebral amyloid angiopathy in the aetiology and immunotherapy of Alzheimer disease. Alzheimers Res. Ther. 1, 6 (2009)
12. Yamada, M.: Predicting cerebral amyloid angiopathy-related intracerebral hemorrhages and other cerebrovascular disorders in Alzheimer's disease. Front Neurol. 3, 64 (2012)
13. Weller, R.O., Preston, S.D., Subash, M., Carare, R.O.: Cerebral amyloid angiopathy in the aetiology and immunotherapy of Alzheimer disease. Alzheimer's Research & Therapy 1, 6 (2009)
14. Thal, D.R., Griffin, W.S., de Vos, R.A., Ghebremedhin, E.: Cerebral amyloid angiopathy and its relationship to Alzheimer's disease. Acta Neuropathologica 115, 599–609 (2008)

# Graph Compression Strategies
# for Instance-Focused Semantic Mining

Xiaowei Jiang[1], Xiang Zhang[2], Feifei Gao[1], Chunan Pu[1], and Peng Wang[2]

[1] College of Software Engineering, Southeast University, Nanjing, China
{xiaowei,ffgao,chunan}@seu.edu.cn
[2] School of Computer Science and Engineering, Southeast University,
Nanjing, China
{x.zhang,pwang}@seu.edu.cn

**Abstract.** Semantic mining is a research area that sprung up in the last decade. With the explosively growth of Linked Data, instance-focused Semantic Mining technologies now face the challenge of mining efficiency. In our observation, graph compression strategies can effectively reduce the redundant or dependent structures in Linked Data, thus can help to improve mining efficiency. In this paper, we first describe Typed Object Graph as a generic data model for instance-focused Semantic Mining; and then we propose two graph compression strategies for Linked Data: Equivalent Compression and Dependent Compression, each of which is demonstrated in specific mining scenarios. Experiments on real Linked Data show that graph compression strategies in Semantic Mining is feasible and effective for reducing the volume of Linked Data to improve mining efficiency.

**Keywords:** Semantic Mining, graph compression, linked data.

## 1 Introduction

Semantic Mining is a research area that sprung up in the last decade. It combines the Semantic Web with data mining, adapting various mining techniques to discover useful information in semantic data. In these years, Semantic Mining has undergone a transition from ontology-driven mining to instance-focused mining, from text mining to graph mining. In [1], for instance, Semantic Mining can be used to classify web documents based on text analysis. While introduced in [2], Semantic Mining techniques are used to discover frequent patterns and semantic associations, basing on an analysis on graph structure.

Lots of graph-based Semantic Mining algorithms have been proposed in various mining scenarios. Some typical scenarios include: analyzing semantic relationships between concepts or instances defined in ontologies, discovering patterns in RDF graphs, importance or popularity assessment of instances in RDF graphs, etc. Most of these algorithms have reasonable computational efficiency only on small datasets. However, as the explosive growth of Linked Data, these algorithms become more and more difficult to be adapted to large-scale Linked Data, which may consist of billion triples.

G. Qi et al. (Eds.): CSWS 2013, CCIS 406, pp. 50–61, 2013.

A possible solution is to divide large-scale Linked Data into suitable size of partitions prior to the process of mining, such as proposed in [3]. This approach is effective for large-scale mining, but is meanwhile complicated. The cost of integrating mining results in partitions may be high, and the approach is also theoretically prone to a loss of mining results. A more simple and intuitive approach is needed. In our observation, there are usually lots of repeated or interdependent structures in Linked Data. Given this characteristic, a structure-based compression can be performed on Linked Data prior to mining process.

In this paper, we propose a framework of two strategies for graph compression to reduce the volume of Linked Data. The first strategy is named Equivalent Compression, which reduces Linked Data by combining repeated structures; the second strategy is named Dependent Compression, which reduces Linked Data by contracting dependent structures. After compression, some graph structures in Linked Data will be combined or contracted into the inner structure of a special instance, which is called a "hypernode" in this paper, and the original graph will be consequently transformed into a relatively smaller "hypergraph". Our work is applicable to instance-focused Semantic Mining tasks, which usually discover instance-related information in Linked Data. Furthermore, different strategies of compression are applicable to different mining scenarios.

This paper is presented as following: a generic graph model for instance-focused Semantic Mining is firstly introduced in section 2. In section 3, a framework of two compression strategies will be described in detail. Two typical mining scenarios using graph compression are well-discussed in section 4 to convince the practicability of the strategies. Finally, experiments are conducted to make a quantitative analysis on how these compression strategies can reduce the volume of Linked Data to improve the efficiency of Semantic Mining.

## 2   Graph Model for Instance-Focused Semantic Mining

In the context of this paper, we refer to Semantic Mining as a mining on graph structures in Linked Data. Furthermore, an instance-focused Semantic Mining is a set of special mining tasks, which focus on discovering instance-related information or knowledge in Linked Data, not schema-related. For example, in [4], a Semantic Mining approach was put forward to find associations among semantic objects on the basis of a pattern-growth-based mining algorithm. As the growth of online Linked Data, instance-focused Semantic Mining has attracted lots of research interest.

Type information of instances is usually important for various mining tasks in instance-focused Semantic Mining. Proposed in [2], Typed Object Graph (TOG in short) is an appropriate graph model for generic-purposed Semantic Mining. TOG is derived from RDF graph by attaching type information to each instance. In the model, each instance has a unique identification and a unique type-attribute. For those instances that are defined to have multiple types, a set of rules are defined to determine their unique type-attribute according to the popularity, importance or universality of their types. A definition of TOG is defined in Definition 1. A real example of TOG is shown in Figure 1. In the example, Tim Berners-Lee is connected to other persons or places.

**Definition 1. (Typed Object Graph):** *Defining quintuple Q as $\langle s, type(s),$ $p, o, type(o) \rangle$ where s, p, o represents the subject, predicate and object of an RDF triples. In Q, s and o should denote instances, which means their rdf:type should not be classes or properties. type(s) and type(o) are the unique types of s and o respectively. Typed Object Graph G is a directed and labeled graph formed by a set of quintuple $G = Q_1, Q_2, \ldots, Q_n$.*

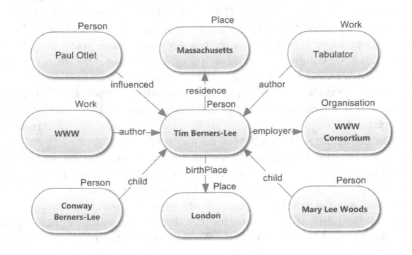

**Fig. 1.** An Example of TOG

## 3    Compression Strategies

The basic idea behind of our compression strategies is to use a single instance to represent a compressible graph structure, which connects a group of highly related instances in TOG. The single representative instance is named as "hypernode" in TOG, which can be an actual instance in TOG, or be a virtual instance standing for the compressed graph structure. The compressed instances are usually topologically close in TOG, and the compressed graph structures are usually repeated or inter-dependent structures. Our compression strategies make use of these structures, and reduce a TOG into a smaller graph.

In this section, a framework of two compression strategies will be discussed. The notion of "Family" of an instance is defined to describe the topological context of an instance in TOG. An Equivalent Compression strategy and a Dependent Compression strategy are separately proposed to reduce repeated structures and inter-dependent structures in TOG respectively.

## 3.1    The Family of an Instance

In Typed Object Graph, the topological context of an instance can be seen as its neighbors and its relations to its neighbors. The neighboring information about an instance is a the local structure around the instance, and usually characterizes its semantics. We use the notion "Family" to stand for the neighboring information.

**Definition 2. (Family of an Instance):** *The "family" of an instance family(i) is a set $F=\langle j,r\rangle|$ j is one of the direct neighbor of i, r is the relation from i to j or j to i.*

## 3.2    Equivalent Compression

In traditional theory of graph compression, one of the feasible approach is based on the similarity between two nodes in a graph. A set of highly similar nodes often leads to repeated and compressible graph structures. The similarity between nodes can be measured from different aspects, such as the node's label, type, degree, or shortest distance to their neighbors, and so on. Motivated by this idea, we use the notion of Family to evaluate the similarity between two instances. The fact that two instances have a very same Family will definitely indicates that they possess a very similar topological position in TOG, which means there are repeated structures in the graph. A virtual hypernode will be created to represent this repeated structure.

Equivalent Compression, namely, is to combine instances whose types are the same and who have an Equivalent Relationship in their Families.

**Definition 3. (Equivalent Relationship):** *Give a TOG $G = \langle V, E\rangle$ derived from an RDF graph, in which V is the set of nodes and E is the set of edges, there is an Equivalent Relationship between instance $i_1$ and instance $i_2$: Equivalent($i_1$, $i_2$) iff 1) $i_1, i_2 \in V$; 2) type($i_1$)=type($i_2$) ; 3) family($i_1$)=family($i_2$).*

**Definition 4. (Equivalent Compression):** *Give a TOG $G=\langle V, E\rangle$, Equivalent Compression is a process of transforming G into another graph: $ECG(G)=\langle V', E'\rangle$ where $V'=V_{hyper} \cup V_{simple}$. $V_{hyper}$ is a set of virtual hypernodes, in which each node represents a set of Equivalent instances; $V_{simple}$ is a set of actual nodes in TOG, in which each node has no Equivalent Relationship to any other nodes in $V_{simple}$.*

As shown in Figure 2, Figure2(a) is a fragment of TOG and Figure2(b) is a corresponding compressed graph. In Figure2(a), both Tom and Mary are Persons, and they both know Jack and Kate. Therefore, Tom and Mary are considered to have Equivalent Relationship and can be combined into a virtual hypernode Tom&Mary. Hypernode is presented in thick line. A created hypernode represents multiple equivalent instances in the original TOG, thus can reduce repeated structure and improve the efficiency of Semantic Mining.

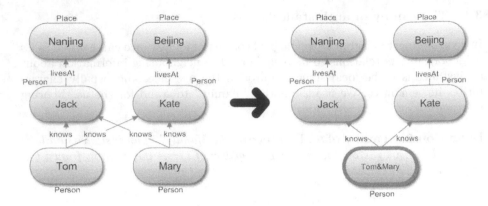

**Fig. 2.** (a)A subgraph of TOG (b)Corresponding ECG

### 3.3  Dependent Compression

Even in a dense TOG, there are still a lot of instances whose family has only one member, which means this kind of instances are not popularly referred by other instances, and their semantics are highly dependent on their only neighbor. They are terminal nodes hanging at the edge of a network formed by interlinked instances. In some Semantic Mining tasks, this kind of dependent structures can be compressed to improve the mining efficiency. The local structure of dependent instances can be contracted into the inner structure of its only neighbor. Different with Equivalent Compression, hypernodes are actual instances in Dependent Compression.

**Definition 5. (Dependent Relationship):** *Give the TOG G=⟨V, E ⟩, there is a Dependent Relationship between instance $i_1$ and $i_2$: Dependent($i_1,i_2$) iff 1) $i_1,i_2 \in V$; 2) $i_2$ is the only instance in family($i_1$) and there is only one edge connecting $i_1$ and $i_2$.*

**Definition 6. (Dependent Compression):** *Give a TOG G=⟨V, E ⟩. Dependent Compression is the process of transforming G into another graph: DCG(G)= ⟨V', E' ⟩, where $V'=V_{hyper} \cup V_{simple}$ .$V_{hyper}$ is a set of hypernodes. Each hypernode contains an actual instance with all instances that have Dependent Relationship to it. $V_{simple}$ is a set of actual instances in TOG, in which there doesn't exist two instance i and j in $V_{simple}$ that satisfy Dependent(i,j) or Dependent(j,i).*

In this strategy, if Dependent($i,j$), instance $i$ will be compressed into $j$, making $j$ a hypernode. The original TOG is compressed in an iterative manner until no dependent instances can be compressed into other instances, and the compression ratio becomes steady.

As shown in Figure 3, the cardinality of family of each instance is first computed. And then, one-neighbor instances ("Tom" and "Mary") are compressed into their only neighbor, which forms the structure on the top of the figure.

The hypernode is presented in thick lines. For lack of space, the inner structures of hypernodes are not presented. The compressed graph can be iteratively compressed, until no Dependent Relationship exists, as the structure on bottom right of the figure, then the compression process comes to the end.

It can be proved that, after the iterative process, the inner structure of hypernodes is usually a tree structure without regards to the direction of quintuples in the tree. Each hypernode in DCG is an actual instance in TOG and is the root of the compressed tree.

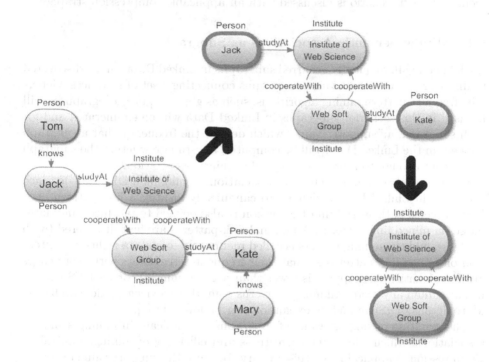

**Fig. 3.** An Example of Iterative Compression of TOG to Corresponding DCG

## 4    Application Scenarios

Our strategies can be applied to instance-focused Semantic Mining to achieve a better mining efficiency. In this section, two typical Semantic Mining tasks will be fully discussed as the application scenarios of our strategies. These scenarios are about the mining of semantic associations. Semantic associations are usually defined as a graph structure representing a group relationship among several instances as defined in [4], or a path structure representing a serial relationship between two instances as defined in [5,6]. Since the complexity of mining semantic associations are usually exponential, they will be inefficient especially in large-scale Linked Data. The mining process often consumes a large amount of time.

The applicability of each compression strategy depends on whether the mining task will make use of the inner structure of hypernodes. For example, since tree structures will be compressed in Dependent Compression, it will not be applicable to the mining tasks that rely on counting the subgraph frequency, such as discovering graph-structured semantic associations in the first application scenario. But Dependent Compression is applicable to discovering path-structured semantic associations, because in mining tasks on paths, the tree structure of hypernodes can be utilized to find shortest path between instances. In following scenarios, each scenario is discussed with an applicable compression strategy.

## 4.1   Mining Semantic Associations as Subgraphs

As defined in [4], frequently occurred subgraphs in Linked Data will be discovered to discover semantic associations as graphs connecting a set of instances. Generally, frequent pattern mining algorithms, such as gSpan [7] or Closegraph [8] will be used in this scenario. Subgraphs in Linked Data will be enumerated and for each subgraph, a "support" value, which defines the frequency that a subgraph appears in the Linked Data, will be computed to estimate whether the subgraph is frequent enough to represent a typical semantic association.

In the process of mining semantic associations, a minimum DFS code for each subgraph in Linked Data is defined to canonically identify a link pattern by a DFS traverse path. A rightmost extension is also defined to produce candidates based on mined link patterns. A minimal link patterns are first discovered (with 0-edges), and the mining process is called recursively to make a rightmost extension on mined link patterns so that their frequent descendants with more edges are found until their support is lower than a given min-sup or its DFS code is not minimum any more. All mined patterns comprise a lexicographic search tree. More details of gSpan and its expansion can be found in [7].

The counting of support value of subgraphs is significant in mining semantic associations, which affects the correctness and efficiency of mining. Equivalent Compression is applicable for this scenario because the support value of a subgraph is easy to compute after compression. In Equivalent Compression, the inner structure of a hypernode is a set of equivalent instances, rather than a tree structure. Therefore, it is not necessary to decompress hypernodes to count the support of subgraph. The only information needed to count the support value in ECG is the cardinality of compressed instances in hypernodes.

When semantic associations are discovered using gSpan or CloseGraph, there is no difference when counting the support value of subgraphs containing only simple nodes after compression. But a modification on gSpan or CloseGraph is needed in the case of counting the support value of subgraphs containing hypernodes: the contribution of the occurrence of a hypernode to the support value is not 1, but the cardinality of the set of compressed instances (saying $k$), because the hypernode represents $k$ equivalent instances, in which each instance contributes one occurrence to the support value.

## 4.2   Mining Semantic Associations as Paths

In some specific Semantic Web applications, such as semantic social network, researchers often focus on discovering relations between two persons, which are usually described as paths between two instances in RDF graph. The closeness of two instances is usually measured by the length of the path. A mining task to discover semantic associations between instances can be transformed into an problem of shortest path discovering.

Dependent Compression can help to improve the mining efficiency of this type of mining tasks. The improvement is based on the idea that shortest path between two nodes in a tree is easier to compute than in a graph. In a graph compressed by Dependent Compression, instances can be classified into three sets - simple instances, hypernodes as the roots of their inner tree structures and other compressed instances in the inner tree structure of hypernodes. $S$, $R$ and $I$ are used to denote each corresponding set of instances.

Naming the shortest path between instance $i$ and $j$ as $SP(i, j)$. $SP(i, j)$ can be computed in separate cases: 1) $SP(i, j)$ can be computed using the general all-pair-shortest-path algorithm when $i, j \in S \cup R$; 2) In the case that $i \in I$, $j \in S \cup R$, finds the root hypernode of $i$ and names it as $k$, $k \in R$, then $SP(i, j) = SP(i, k) \mid SP(k, j)$; 3) In the case that $i, j \in I$, finds the root hypernodes of $i$ and $j$ and names it as $k, l$ respectively, then $SP(i, j) = SP(i, k) + SP(k, l) + SP(l, j)$.

# 5   Experiment

To validate the effectiveness of two compression strategies, a set of experiments are conducted on five online Linked Data. In this section, we first describe the dataset, and then discuss the experiments and results.

## 5.1   DataSet

To validate the effectiveness of the compression strategies, five online Linked Data on different topics are selected as our dataset for evaluation. A summary of each dataset is given as following:

(i)   **DBpedia**, which is a widely-used Linked Data on structured information extracted from Wikipedia.
(ii)   **LinkedMDB**, which is a famous Linked Data on movies.
(iii)   **SwetoDblp**, which is an ontology focused on bibliography data of publications from DBLP with additions that include affiliations, universities, and publishers.
(iv)   **Jamendo**, which is a large Linked Data of Creative Commons licensed music, based in France. These datasets are diverse in topics, and are rather large in volume, which makes them difficult for Semantic Mining tasks.
(v)   **John Peel sessions**, which is published by DBTune.org. It is a music-related repository on the Semantic Web, containing Linked Data of BBC John Peel sessions.

All the datasets have RDF dumps in their corresponding websites, and can be accessed through a portal of W3C DataSetRDFDumps[1] The statistical information of these datasets are given in Table 1.:

Table 1. The Statistical Information of Datasets

| Dataset | #instancs | #quintuple | Ave.Degree |
|---|---|---|---|
| DBpedia | 1664061 | 6014163 | 7.228 |
| LinkedMDB | 602796 | 1210921 | 4.0177 |
| SwetoDblp | 544678 | 627753 | 2.305 |
| Jamendo | 281468 | 373494 | 2.654 |
| JohnPeel | 71284 | 100403 | 1.408 |

## 5.2    Evaluation on Compression

The compression ratio $CR$ is a conventional parameter to measure the performance of compression. It describes the relative volume of data after compression comparing to the volume of data before compression. In graph compression, $CR$ can be defined either by compression ratio on number of edges or by compression ratio on number of nodes. In this paper, we define $CR_q$ and $CR_i$ in Equation 2-3 to represent the compression ratio on quintuples or instances in Linked Data respectively. A low compression ratio indicates that a large proportion of quintuples or instances in TOG can be compressed according to compression strategy.

$$CR_q = \frac{\#quintuple\,after\,compression}{\#quintuple\,before\,compression} \qquad (1)$$

$$CR_i = \frac{\#instance\,after\,compression}{\#instance\,before\,compression} \qquad (2)$$

Both Equivalent and Dependent Compression Strategy are performed on each dataset, and the compression ratios are shown in Table 2 and Table 3 respectively. In the tables, $\#q$ and $\#i$ indicate the number of quintuples and instances in TOG. $-q$ and $-i$ indicate the number of compressed number of quintuples and instances in ECG and DCG.

From the results in Table 2, it is observable that $-q$ is normally higher than $-i$, because a compression of a set of equivalent instances usually leads to a compression of a larger set of equivalent quintuples in the families of these instances. In our observation, the graph structures in DBpedia, LinkedMDB and JohnPeel are non-redundant, which means few repeated structures can be discovered by Equivalent Compression. On these datasets, our strategies have limited performance, with both $CR_q$ and $CR_i$ are over or almost over 90%. Especially for John-Peel, only one pair of equivalent instances can be found. However, Equivalent Compression have sound performance on the other two datasets. For SwetoDblp,

---

[1] W3C DataSet RDF Dumps: http://www.w3.org/wiki/DataSetRDFDumps.

**Table 2.** Compression Ratio of Equivalent Compression

| Dataset | Before Compression | | After Compression | | $-q$ | $-i$ | $CR_q$ | $CR_i$ |
|---|---|---|---|---|---|---|---|---|
| | #q | #i | #q | #i | | | | |
| DBpedia | 6,014,163 | 1,664,061 | 5,877,935 | 1,555,424 | 136,228 | 108,637 | 0.977 | 0.935 |
| LinkedMDB | 1,210,921 | 602,796 | 1,097,710 | 537,653 | 113,211 | 65,143 | 0.907 | 0.892 |
| SwetoDblp | 627,753 | 544,678 | 146,730 | 64,038 | 481,023 | 480,640 | 0.234 | 0.118 |
| Jamendo | 373,494 | 281,468 | 306,328 | 214,568 | 67,166 | 66,900 | 0.820 | 0.762 |
| JohnPeel | 100,403 | 71,284 | 100,401 | 71,283 | 2 | 1 | 1.000 | 1.000 |

**Table 3.** Compression Ratio of Dependent Compression

| Dataset | Before Compression | | After Compression | | $-q$ | $-i$ | $CR_q$ | $CR_i$ |
|---|---|---|---|---|---|---|---|---|
| | #q | #i | #q | #i | | | | |
| DBpedia | 6,014,163 | 1,664,061 | 5,724,074 | 1,373,972 | 290,089 | 290,089 | 0.952 | 0.826 |
| LinkedMDB | 1,210,921 | 602,796 | 1,068,716 | 460,591 | 142,205 | 142,205 | 0.883 | 0.764 |
| SwetoDblp | 627,753 | 544,678 | 115,372 | 32,297 | 512,381 | 512,381 | 0.184 | 0.059 |
| Jamendo | 373,494 | 281,468 | 150,289 | 58,263 | 223,205 | 223,205 | 0.402 | 0.207 |
| JohnPeel | 100,403 | 71,284 | 59,787 | 30,668 | 40,616 | 40,616 | 0.595 | 0.430 |

surprisingly almost 90% of instances with 80% of quintuples are equivalent and can be compressed.

Shown in Table 3, Dependent Compression have a better compression ratio comparing to Equivalent Compression. All the ratios, except $CR_q$ on DBpedia, are less than 90%. Dependent Compression also has the best performance on SwetoDblp, in which $CR_i$ can even reach to 6%. That indicates SwetoDblp consists of too many repeated and meanwhile dependent structures. Another observable fact is that the $-q$ is just the same with $-i$ for each dataset after Dependent Compression. This is because a compression of a dependent instance always leads to a simultaneous compression of the quintuple connecting it to its only neighbor. In other words, the number of instances and the number of quintuples in the inner structures of hypernodes in DCG are always the same.

# 6 Related Work

The goal of graph compression for graph mining is to reduce the volume of graph and to achieve a reasonable mining efficiency when the graph is large. Some approaches have been proposed for graph compression. In [12], Chen proposed a graph mining algorithm via randomized graph summaries. For each graph in the graph set, a summary is built and the shrunk graph is used for mining, which decreases the embedding enumeration cost. However, after a graph summarization, the algorithm may suffers from a loss of patterns. Thus randomized summaries are generated and the mining process is repeated for multiple rounds for a minimum loss of patterns. The works presented in [13] and [14] are closely related to ours. These paper independently proposed to construct graph summaries of unweighted graphs by grouping similar nodes and edges to supernodes and superedges. The difference between their works and ours lies in two aspects: first, their works are not working on Linked Data, type information is undefined for the nodes in their graph models; second, their works considered the compression of repeated structures, but not considered compression of dependent structures.

Besides these works, some other solutions on graph compression have been proposed. Toivonen proposed in [15] a solution to compress graph by node and edge mergers. Toivonen also introduce another solution in [16], which is also known as graph simplification. Nodes and edges in weighted graphs are grouped to supernodes and superedges. The supernodes and superedges are selected to minimize approximation errors and meanwhile maximize the amount of compression. Both of the works provides approach to approximate compression, and thus are both lossy. They are quite different with our work, which works on non-weighted graph and is lossless in the process of compression and decompression.

# 7    Conclusion and Future Work

The research on Semantic Mining is now facing the challenge of contradiction between the growing volume of Linked Data and the complexity of mining algorithms. The approach of graph compression can effectively improve the efficiency of instance-focused Semantic Mining by simplifying the graph structure of large-scale Linked Data. In this paper, a set of two strategies are proposed for graph compression on a generic graph model for instance-focused Semantic Mining. Repeated and Dependent structures in TOG are compressed by Equivalent and Dependent Compression respectively. Two typical scenarios are discussed to illustrate the applicability of each strategy. Experiments on real datasets show that our approach is feasible to reducing repeated and dependent structures in TOG, and practically improves mining efficiency in typical application scenarios.

The combination of graph compression and graph partitioning will be considered in our future work. Indicated by experimental results, graph compression strategies have limited performance on very densely-connected TOG. For mining on this kind of TOG, graph partitioning will help to divide TOG into minable blocks. The combination of graph compression and partitioning is expected to provide a complete solution to large-scale Semantic Mining.

**Acknowledgements.** The work is supported by the NSFC under Grant 61003055, 61003156, and by NSF of Jiangsu Province under Grant BK2011335. We would like to thank Lei Wu and Yu Guo for their valuable suggestions and work on related experiments.

# References

1. Svatopluk, F., Ivan, J.: Semantic Mining of Web Documents. In: Proceedings of International Conference on Computer Systems and Technologies, pp. 21–26 (2005)
2. Zhang, X., Zhao, C., Wang, P., Zhou, F.: Mining Link Patterns in Linked Data. In: Gao, H., Lim, L., Wang, W., Li, C., Chen, L. (eds.) WAIM 2012. LNCS, vol. 7418, pp. 83–94. Springer, Heidelberg (2012)
3. Zhao, C.F., Zhang, X., Wang, P.: A Label-based Partitioning Strategy for Mining Link Patterns. In: Proceedings of 7th International Conference on Knowledge, Information and Creativity Support Systems, pp. 203–206 (2012)
4. Jiang, X.W., Zhang, X., Gui, W., Gao, F.F., Wang, P., Zhou, F.B.: Summarizing Semantic Associations Based on Focused Association Graph. In: Proceedings of the 8th International Comference, pp. 564–576 (2012)
5. Anyanwu, K., Sheth, A.: p-Queries: Enabling Querying for Semantic Associations on the Semantic Web. In: Proceedings of the 12th International World Wide Web Conference, pp. 690–699 (2003)
6. Sheth, A., Aleman-Meza, B., Arpina, I.B., et al.: Semantic Association Identification and Knowledge Discovery for National Security Applications. Journal of Database Management 16(1), 33–53 (2005)
7. Yan, X., Han, J.W.: gSpan: Graph-based Substructure Pattern Mining. In: Proceedings of the IEEE International Conference on Data Mining, pp. 721–724 (2002)
8. Yan, X., Han, J.W.: CloseGraph: Mining Closed Frequent Graph Patterns. In: Proceedings of the 9th ACM SIGKDD International Conference on Knowledge Discovery and Data Mining, pp. 286–295 (2003)
9. Hage, P., Harary, F.: Eccentricity and Centrality in Networks. Social Networks 17, 57–63 (1995)
10. Page, L., Brin, S., Motwani, R., Winograd, T.: The PageRank Citation Ranking: Bringing Order to the Web. Technical Report, Stanford University (1998)
11. Kleinberg, J.: Authoritative Sources in a Hyperlinked Environment. In: Proceedings of the 9th ACM SIAM Symposium on Discrete Algorithms, pp. 668–677 (1998)
12. Chen, C., Lin, C.X., Fredrikson, M., Christodorescu, M., Yan, X.F., Han, J.W.: Mining Graph Patterns Efficiently via Randomized Summaries. In: Proceedings of the 35th International Conference on Very Large Data Bases, pp. 742–753 (2009)
13. Navlakha, S., Rastogi, R., Shrivastava, N.: Graph Summarization with Bounded Error. In: Proceedings of the 2008 ACM SIGMOD International Conference on Management of Data, pp. 419–432 (2008)
14. Tian, Y., Hankins, R., Patel, J.: Efficient Aggregation for Graph Summarization. In: Proceedings of the 2008 ACM SIGMOD International Conference on Management of Data, pp. 567–580 (2008)
15. Toivonen, H., Zhou, F., Hartikainen, A., Hinkka, A.: Network Compression by Node and Edge Mergers. In: Berthold, M.R. (ed.) Bisociative Knowledge Discovery. LNCS, vol. 7250, pp. 199–217. Springer, Heidelberg (2012)
16. Toivonen, H., Zhou, F., Hartikainen, A., Hinkka, A.: Compression of Weighted Graphs. In: Proceedings of the 17th ACM SIGKDD International Conference on Knowledge Discovery and Data Mining, pp. 965–973 (2011)

# Searching Semantic Associations
# Based on Virtual Document

Chen Wang[1], Xiang Zhang[2], Yongtao Lv[1], Li Ji[1], and Peng Wang[2]

[1] College of Software Engineering, Southeast University, Nanjing, China
{onechen,yongtaolv,jillier}@seu.edu.cn
[2] School of Computer Science and Engineering, Southeast University,
Nanjing, China
{x.zhang,pwang}@seu.edu.cn

**Abstract.** As the explosive growth of online linked data, enormous RDF triples are produced every minute in various fields such as health, transportation, chemical, etc. There is an urgent need for an approach to finding and searching semantic association from massive data. However, the complex graph structure of the semantic association brings a great barrier to the process of searching. Transforming the complex graph into text-based structure is a better idea. To characterize the semantics of each association, a virtual document of each association is built with the help of a neighboring operation. A searching model of virtual documents of associations and a ranking schema are also discussed in this paper. Experiments show that our approach is feasible and efficient.

**Keywords:** linked data, semantic association, link pattern, virtual document.

## 1 Introduction

With the explosive growth of semantic web in this decade, there is an exponential growth in the scale of online linked data. Every day, every minute, enormous amount of linked data are produced by social communities, companies, and even by end-users in various fields. Linked data provide a good practice for connecting and sharing semantic objects by URI and RDF. An important knowledge we can discovered in linked data is the explicit or hidden relationship between or among semantic objects, which is terminologically named as semantic associations. An early statement of semantic association can be found in [1], in which semantic associations are connections between two objects, and are represented as semantic paths in RDF graph. Discussions of mining semantic associations in linked data have lasted for near ten years. However few studies have been done in aspects of searching semantic associations from the linked data.

Derived from linked data, there are frequent styles of interlinked objects called linked pattern and lots of semantic associations are discovered from the linked patterns. Both of link pattern and semantic association are essentially structured as a complex graph. Referred in Mining Link pattern in linked data, one pattern

G. Qi et al. (Eds.): CSWS 2013, CCIS 406, pp. 62–75, 2013.

or association may contain at most 23 closely connected edges. However, searching from the complex graph network is always a troublesome problem in the field of information retrieval. The text-based approach to finding and searching pattern or association is very necessary and worthwhile. The second, the result of transforming graph-based pattern or association is normally short text-based structure, which may cause the loss of information and the decrease of the recall rate. Then we consider merging the neighboring information into text derived from each pattern and association because the neighboring information also describes the semantics for its own in some extent. In this paper, we borrowed a notion of virtual document referred in [6] and have it modified to adapt to the searching process of linked patterns and semantic associations.

## 2   System Architecture

In this section, we give an overview of the architecture of our searching system. As shown in Figure 1, the input of the system is various sources of linked data. The goal of the system is to enable a keyword-based search of link patterns and semantic associations. Like the architecture of traditional information retrieval system, the users submit their keyword queries to the system, and hit results are selected from the association or pattern index and then shown in accordance with a text-based ranking scheme. But the difference is, in our system, the text indexes are built from the virtual documents of mined semantic associations and link patterns. That means the text index not only embodies the linguistic information of these associations and patterns, but also reflects the graph structure of them.

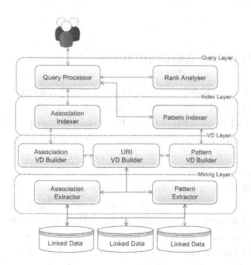

**Fig. 1.** The architecture of searching system of semantic association

**Linked Data**: The input of system is various sources of linked data in RDF/XML or OWL formats.

**Association/Pattern Extractor**: This module discovers link patterns by applying a frequent pattern mining algorithm, and then semantic associations are extracted in an instantiation process by connecting multiple objects.

**URI VD Builder**: This module analyzes the information of each URI reference occurred in linked data. The virtual document of each URI includes both the local descriptions and its neighboring information.

**Association/Pattern VD Builder**: This module collects virtual document of each URI occurred in the association or pattern and then merges them together to generate a complete virtual document.

**Association/Pattern Indexer**: In this module, Lucene is used to create an efficient text-based index for the virtual documents of semantic associations and link patterns separately.

**Rank Analyzer**: In this module, we combine the query-relevant ranking schema TF-IDF and a PageRank-based link analysis ranking scheme to calculate the rank value of each association or pattern.

**Query Processor**: This module works as the query interfaces to end users. Receiving keywords submitted by user, it sends the query results generated by searching methods to Rank Analyzer. Then it picks up top-k sorted results to user.

## 3    Mining Semantic Associations

The notion of semantic association is traditionally defined as paths between two objects. Declared in [1], two objects are semantically associated if they are semantically connected by a semantic path. Anyanwu proposed a notion of property sequence in [2] to define the semantic association between two objects. In [3], Kochut used Defined Directionality Path to characterize semantic associations.

In real linked data, two semantic associations are shown in Figure 2(a). One association represents a relation among three objects in a paper presentation event: The author Aleman-Meza presented a research paper on query result ranking, which is published in the conference ISWC2003. The other association is similar, representing another relation among three other objects: The author Rocha presented his paper on semantic search in the conference WWW2004. Both associations have a common pattern among Author, Publication and Conference as shown in Figure 2(b). This pattern is frequently occurred in linked data, which indicates that the pattern represents pragmatically typical associations among different types of objects. Such patterns are called link pattern.

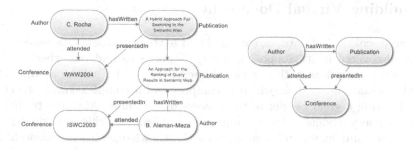

**Fig. 2.** (a) An example of semantic associations of six objects (b)corresponding link pattern

A full discussion of link pattern had been given in [4]. Semantic associations can be seen as instantiation of frequently occurred link patterns.

Link patterns cannot be directly mined from the RDF graph of linked data, because object types are core elements in link patterns. However, in RDF graphs, object types are implicit and can only be determined by reasoning according to RDF semantics. We have proposed Typed Object Graph (TOG in short) in [4] as the graph model for mining link patterns. A TOG is derived from an RDF graph in linked data, in which each triple is extended to a link quintuple, additionally containing the types of the subject and the object in the original triple.

Our mining approach of link patterns follows the idea of pattern-growth-based frequent pattern mining. We adopt gSpan [5] as the mining algorithm. Link patterns provide a schema-level template for mining semantic associations. Each discovered semantic association is certain to be frequent and typical in linked data. Practically, link patterns and semantic associations are mined simultaneously during each iteration of gSpan. That means all semantic associations will be mined and the number of semantic associations equals to the summation of total supports of all link patterns. The notion of semantic association is defined as following:

**Definition 1.** *(Instantiation of Link Pattern): Given an linked data d, and RDF Triple $g = \langle V(g), E(g) \rangle$ derived from d, a discovered link pattern $p = \langle V(p), E(p) \rangle$ can be instantiated by g, iff:*
*(1)$\forall u \in V(g)$, u is an object.*
*(2)$\forall v \in V(p)$, $\exists v' \in V(g)$ and type(v', d)=v*
*(3)$\forall (u,v) \in E(p)$, $\exists (u', v') \in E(g)$ and type(u', d)=u and type(v', d)=v.*
*(4)g is minimal, type(v) is the type of vertex v.*

**Definition 2.** *(Semantic Association): Given a set of objects $O = o_1, o_2, ..., o_i$ in a linked data, a semantic association sa(O,E) is an RDF graph derived from d, iff sa(O,E) is an instantiation of a discovered link pattern in d. O is the vertex set of sa(O,E), and E is the edge set.*

## 4    Building Virtual Document

For a large-scale linked data, an extremely large volume of semantic associations can be discovered in it. Supposing a user is going to find out whether there are certain associations with a keyword query, the results will be a large set of related associations, each of which is composed of a complex graph structure. This will bring a great barrier to the process of searching. Also, the traditional structural query models and algorithms are inapplicable in this scenario, because they are too complex for end users without the knowledge on semantic web. It is necessary for us to find a method to transform the complex graph into text-based structure, which will be easier for user to understand and search.

Therefore, we borrowed the notion called virtual document referred in [6] to store the information of each object derived from the node it represents. However, the virtual document of each semantic association or link pattern could be very short, and short document search will usually leads to a relatively low recall. We use a neighboring operation in constructing the virtual document, which extends a virtual document of an association or pattern by adding virtual document of other related associations or patterns. The virtual document is bag of words derived or inferred from the context of the document defined in the IR model in this paper.

In this section, we will introduce the process of building virtual document for URI, link model and semantic association. And some important methods, definitions, symbols referred in this process are listed as follow.

$\bigcup$:the operator stands for merging several sets of words together(repeating words are unadmitted).

$\uplus$:the operator stands for merging several bags of words together(repeating words are admitted).

Briefly, for a literal node, the description is a collection of words derived from the literal itself; for a URIref(refer to Definition 1), it is a collection of words extracted from the local name, rdfs:label(s), rdfs:comment(s) and other possible annotations.

**Definition 3.** *(Description of URIrefs):Given a URIref e, the desciption of e is a collection of words defined by (1):*

$$des(e) = \uplus \{LN(e), RL(e), RC(e), OA(e)\} \tag{1}$$

Here, we use $LN(e)$ to represent the bag of words in the local name of $e$, $RL(e)$ to represent the bag of words in the rdfs:label of $e$, $RC(e)$ to represent the bag of the rdfs:comment of $e$ and $OA(e)$ to represent the bag of the other possible annotations.

To include descriptions of neighbors in virtual documents, we use three neighboring operations to capture different kinds of neighbors. For a URIref denoted by $e$, we use $N_1(e)$ to denote the other nodes that occur in triples which contain $e$, and we called it the direct neighboring operation(one-step neighboring). $N_2(e)$ is used to denote $N_1(e)$ adding the neighbors of the nodes derived from $N_1(e)$, and we called it two-step neighboring. K-step neighboring for $N_k(e)$ is analogous The formal definition is given below.

**Definition 4.** *(K-step Neighbors of URIrefs): Let $u$ be a URIref, $t$ be an RDF triple:$t=\langle s(t), p(t), o(t)\rangle$. The K-step neighbors are written as $N_k(u)$, $N_1(u)$ is used to express the direct neighbors and $N_2(u)$ is defined for two step neighbors. Their definitions are given below.*

$$N_1(u) = \bigcup_{s(t)=u} \{p(t), o(t)\} \;\bigcup_{p(t)=u} \{s(t), o(t)\} \;\bigcup_{obj(t)=u} \{s(t), p(t)\} \qquad (2)$$

$$N_2(u) = \bigcup_{u' \in N_1(u)} \{N_1(u), N_1(u')\} \qquad (3)$$

$$N_k(u) = \bigcup_{\substack{e' \in N_{k-1}(u) \\ 0 < i < k}} \{N_i(u), N_1 u'\} \qquad (4)$$

Based on the description formulations and the neighboring operations, we define the virtual document of URIrefs as follows:

**Definition 5.** *(Virtual Documents of URIrefs): Given a URIref $u$. The virtual document of $u$, written as k-step $VD_k(u)$, is defined by (5):*

$$VD_k(u) = \biguplus_{u' \in N_k(u)} \{des(u), des(u')\} \qquad (5)$$

We can easily find that link model and semantic association consist of several nodes and edges, each has its own virtual document, so the virtual document of the link model or semantic association is a collection of each small virtual document.

**Definition 6.** *(Virtual Documents of Link Pattern): Given a Link Pattern $p = \langle V(p), E(p)\rangle$. The k-step virtual document of $p$, written as $VD_k(p)$, is defined by (6).*

$$VD_k(p) = \biguplus_{v' \in V(p)} VD_k(v') \biguplus_{e' \in E(p)} VD_k(e') \qquad (6)$$

**Definition 7.** *(Virtual Documents of Semantic Associations): Given a Semantic Association $sa = \langle O, E\rangle$. The virtual document of $p$, written as $VD_k(sa)$, is defined by (7).*

$$VD_k(sa) = \biguplus_{o' \in O} VD_k(o') \biguplus_{e' \in E} VD_k(e') \qquad (7)$$

(A sample example): To demonstrate the construction of virtual documents, we extract author-conference-publication association from the graph structure shown in Fig 2(a). To simplify the expression, we excludes the built-ins provided by RDF and just use a word to express URIref. There are six URIrefs in this association(we define it as $sa$). According to definition 3, we can easily get the description of each URIrefs, such as:

des(WWW2004)={"WWW2004"}
des(A Hybird Approcah For Searching in the semantic web)={"A", "Hybird", "Approach", "For", "Searching", "in", "the", "semantic", "web" }

Then acoording to definition 4 and 5, K-step virtual document of $sa$ is constructed as followed:

$VD_0(sa)$={"C.Rocha", "hasWritten", "attended", "presentedIn", "A", "Hybird", "For", "Searching", "in", "the", "Semantic", "Web", "WWW2004"}
$VD_1(sa)$={"C.Rocha", "hasWritten", "attended", "presentedIn", "A", "Hybird", "For", "Searching", "in", "the", "Semantic", "Web", "WWW2004", "an", "Approach", "for", "the", "Ranking", "of", "Query", "Results", "in", "Semantic", "Web"}
$VD_2(sa)$={"C.Rocha", "hasWritten", "attended", "presentedIn", "A", "Hybird", "For", "Searching", "in", "the", "Semantic", "Web", "WWW2004", "an", "Approach", "for", "the", "Ranking", "of", "Query", "Results", "in", "Semantic", "Web", "presentedIn", "ISWC2003", "hasWritten", "B.Aleman-Meza" }

Supposing a user eager for the association $sa$(C.Rochas paper in WWW2004): for 0-step virtual document, we can only use the keyword appeared in $sa$ to hit this association. But from the graph structure shown in Fig 2(a), we can easily find that Aleman-Meza also presented a research paper cited by C.Rochas paper in WWW2004. The neighboring information like this can be included in multistep virtual document. For 1-step virtual document, user can use keyword like "Ranking" and "Query" appeared in the cited paper to hit $sa$. For 2-step virtual document, user also can use keyword "Aleman"(the author of the cited paper) to hit $sa$.

## 5   Searching Semantic Associations

### 5.1   Searching Model

Like the traditional information retrieval system, we use IR model as our searching model. However, it is essential to modify the IR model according to the query of user. Then we put forward the notion of the association searching model and pattern searching model. The definitions of them are given as followings:

**Definition 8. *(Association Searching model):*** $\langle A, Q, F, R(q_i, a_j) \rangle$: A is a collection of all the virtual doucment of semantic associations. Q is a collcetion of all the reachable assocaiton queries. F is a framework between A and Q. R(q_i,a_j) is the ranking value provided by ranking fuction.

**Definition 9. (Pattern Searching model):** $\langle P, Q, F, R(q_i, a_j)\rangle$: P is a collection of all the virtual doucment of semantic associations. Q is a collcetion of all the reachable assocaiton queries. F is a framework between P and Q. R(q_i,a_j) is the ranking value provided by ranking fuction.

## 5.2 Ranking Schema

As the key word inserted by user, the result we provide may be large. It's necessary for us to adopt a ranking method to make the searching results more friendlier. Firstly, we need to give dynamic ranking according to the text-relevance. Since our virtual document is constructed by bag model(not set model), there are a lot of repeating words in our virtual document and the traditional ranking method TF-IDF is a good choice. Secondly, considering the links in the data graph, we use the PageRank to finish the static ranking, which has no relation with query.

Next, we will give the detailed introduction of the ranking method. Here, $d_j$ is a substitute of $a_j$ or $p_j$ refered in searching models. We set two parameters $\alpha$ and $\beta$ for these two ranking algorithms and $\alpha + \beta = 1$.

$$R(q_i, d_j) = \alpha \times \frac{TF(q_i, d_j) \times IDF(q_i)}{Max(TF(q, d) \times IDF(q))} + \beta \times \frac{PR(d_j)}{Max(PR(d))} \qquad (8)$$

TF($q_i$,$d_j$) is the frequency of the word $q_i$ occurred in virtual document $d_i$.

$$IDF(q_i) = log\frac{N_{all}}{DF(q_i)} \qquad (9)$$

$N_{all}$ is total number of documents DF($q_i$) is the number of documents in which the word $q_i$ occurred at least once time.

$$PR(d_j) = \frac{1-d}{N} + d\sum_{d_j} \frac{Sim(d_i, d_j) \times PR(d_j)}{\sum_{d_k} Sim(d_j, d_k)} \qquad (10)$$

We can easily extract associations or patterns from the RDF graph, then the PR value of each association or pattern can be calculated. Here, $d_j$ is associations or patterns linked to $d_i$(have at least one common node and edge) and $d_k$ is association or pattern linked to $d_j$. Considering that the affinity of different associations or patterns depend on the number of common node and edges, we define Sim to express their similarities:

$$Sim(d_j, d') = \frac{\sum_{e \in E(d_j) \cap E(d')} e + \sum_{v \in V(d_j) \cap V(d')} v}{\sum_{e \in E(d_j) \cup E(d')} e + \sum_{v \in V(d_j) \cup V(d')} v} \qquad (11)$$

# 6   Evaluation

In this section, we will present the experimental results to perform the evaluation on the process of retrieving. Several experiments are designed to evaluate our approaches. In our experiments, we mainly discuss the space of virtual documents in different levels and comparison in sorting results.

## 6.1   Dataset

The testing data is deprived from the real DBpedia on the Internet. The entire DBpedia dataset contains a completely extracted data from Wikipedia. As of June 2012, the DBpedia 3.8 is released. The dataset describes more than 3.64 million objects using over 1 billion triples. Majority type of objects in DBpedia includes persons, places, music albums, films, video games, organizations, species, diseases and so on. DBpedia has a broad scope of objects covering different areas of human knowledge, and is widely used for the research of semantic knowledge management or semantic search.

In our experiment, we extract objects from the datasets, including 19,170,031 URIs, 617,035 semantic associations and 1054 patterns.

## 6.2   Evaluation on Levels of Virtual Documents

We have performed experiments on the correlation of virtual documents and query result, the average neighbors, time and space consumption with different levels of the virtual documents for semantic associations or link patterns, and a complete evaluation on space and time consumption when the scale of data grows.

We randomly select a set of URIs occurred in different associations or patterns and measure the neighbors of each URI included in virtual document when the level of virtual document increases. Then the average amount of each URI

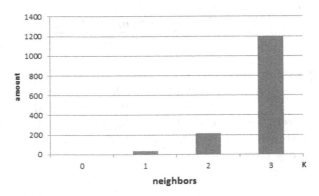

**Fig. 3.** Correlation of the number of neighbors for each URI and level of virtual document

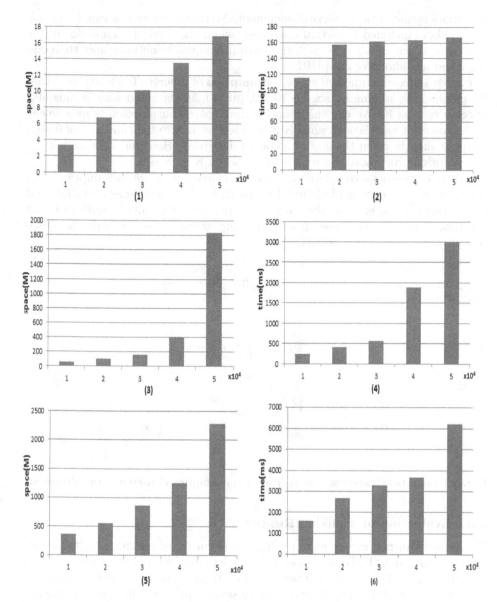

**Fig. 4.** (1)Average space consumption of 0-step virtual doucment (2)Average time consumption of 0-step virtual doucment (3)Average space consumption of 1-step virtual doucment (4)Average time consumption of 1-step virtual doucment (5)Average space consumption of 2-step virtual doucment (6)Average time consumption of 2-step virtual doucment

is calculated, shown in Figure 3. Here, K represents the level of virtual document. Apparently, 0-step virtual documents don't include neighbors. When K is chosen as 2, each URI may have an average size of more than two hundred neighbors included in the virtual document. Moreover, there is a rapid increase of neighbors included in virtual document when the level of virtual document increases. When we select K as 3, the virtual document contains more than one thousand neighbors for each URI.

Figure 4 presents time and space consumption of different levels of the virtual documents. We randomly pick up 10, 20, 30, 40, 50 thousand associations and select 0 to 2 as K. We can easily notice that space consumption is near linear with growth of associations when K is selected as 0. Also the time cost of 0-step increase slightly when the size of associations grows. However, time and space consumption grow more and more rapidly when K increases.

Figure 5 presents a complete evaluation on time and space consumption with the increase of the steps of virtual document. We randomly select 10 thousand associations or patterns to take down the experimental results. It indicates that the time consumption is near linear with the growth of steps while the space consumption is near exponential.

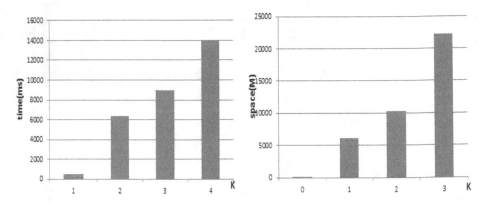

**Fig. 5.** Complete evluation on time and space consumption of K-step virtual document

### 6.3    Evaluation on Ranking Results

We have performed experiments testing the accuracy of ranking results. In our work, we combine TF-IDF and pagerank to calculate the ranking value referred in chapter 5. We select the ratio of two parameters $(\alpha/\beta)$ as 1 to 9 and 0.1 to 0.9 to provide the top 20 records. Then we invite several people to help us test the accuracy of ranking result by different parameter. The thing people need to do is to pick up the useful associations from our ranking results. According to the response of these people, we calculate the average accurate ratio for different parameters shown in Figure 6. It is observable that when $\alpha$ is larger than $\beta$, the average accuracy is comparatively high, which indicates the query-relevent ranking schema TF-IDF should be more important than query-irrelevent PageRank in the overall ranking schema. However, when TF-IDF is quite dominant in the overall ranking schema, such as when ratio is over 6, the accuracy of ranking results will decline. The optimal ranking result occurs when the ratio is setting to 6.

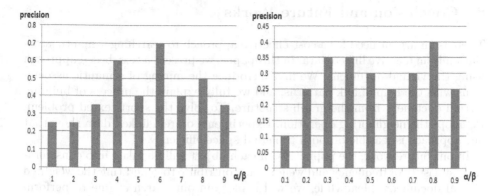

**Fig. 6.** Correlation of the accuracy of ranking results and $\alpha/\beta$

# 7  Related Work

Searching semantic associations is a novel research topic. Our approach is based on text-searching. There are some related works in category of text-searching. Defined by Le [6], ASCO makes use of local descriptions but not neighboring information. It can be considered as the simple virtual document(0-step) in our prospective. Keyword selecting is adopted in the approach by Hu [8], making use of the local description of instances and subclasses. In [7], the classified documents are assumed to be existing beforehand. In GLUE [9], the information contents in the training set for Name Learner and Content Learner are full names and textual contents of instances, respectively. These information contents are generated at run-time and can be seen as virtual documents of instances in our perspective. However, most of their work pay more attention to ontology matching. In this paper, we construct virtual document for semantic searching. Typical applications of text-based searching in searching engine help user quickly and simply find the information needed.

Researches of search in the graph data are also closely related to our work. Path-based indexing approaches introduced by C.A.James [10] are proposed to support queries on general graphs. In [11], Haoliang Jiang introduce a novel sequencing method to capture the semantics of the graph data, which finds meaningful components in graph structures and use them as the most basic unit in sequencing. GraphGrep [12] represents the state-of-the-art technique of path-based graph indexing. For each graph, it enumerates all paths up to length in the graph, and records the number of occurrences of each path. GIndex [13] uses frequent graph patterns instead of paths as index features. Using frequent graph patterns reduces index space and improves the filtering rate. All these work are contributed for graph searching based on graph structure. In our paper, we transform the complex graph into text-based structure, which may be much easier for users to understand.

## 8    Conclusion and Future Works

There is an urgent need for a cost-effective approach to searching semantic association from massive linked data. In this paper, we present a text-based approach using the virtual document. We first introduce the mining of semantic associations with the help of link patterns, and we fully explain the process of building virtual document from the graph structure. To solve the short-texted problem, we adopt the neighboring operation. Experiments on real linked data show that our approach is feasible on both time and space efficiency.

In our future work, the popularity of each object will be taken into consideration, statistics of frequency and weight will contribute to generating the weighted virtual document. Meanwhile, we will explore in our search engine to perform more experiments testing the accuracy of searching results and make it more user-friendly.

**Acknowledgements.** The work is supported by the NSFC under Grant 61003055, 61003156, and by NSF of Jiangsu Province under Grant BK2011335. We would like to thank Xing Li for his valuable suggestions and work on related experiments.

## References

1. Aleman-Meza, B., Halaschek-Wiener, C., Arpinar, I.B., Sheth, A.P.: Context-aware Semantic Association Ranking. In: Proceedings of the 1st International Workshop on Semantic Web and Databases, pp. 33–50 (2003)
2. Anyanwu, K., Sheth, A.: p-Queries: Enabling Querying for Semantic Associations on the Semantic Web. In: Proceedings of the 12th International World Wide Web Conference, pp. 690–699 (2003)
3. Kochut, K.J., Janik, M.: SPARQLeR: Extended Sparql for Semantic Association Discovery. In: Proceedings of the 4th European Conference on Semantic Web, pp. 145–159 (2007)
4. Zhang, X., Zhao, C., Wang, P., Zhou, F.: Mining Link Patterns in Linked Data. In: Gao, H., Lim, L., Wang, W., Li, C., Chen, L. (eds.) WAIM 2012. LNCS, vol. 7418, pp. 83–94. Springer, Heidelberg (2012)
5. Yan, X., Han, J.W.: gSpan: Graph-based Substructure Pattern Mining. In: Proceedings of the 2002 IEEE International Conference on Data Mining, pp. 721–724 (2002)
6. Le, B.T., Dieng-Kuntz, R., Gandon, F.: On Ontology Matching Problems - for Building a Corporate Semantic Web in a Multi-Communities Organization. In: Proceedings of the 2004 International Conference on Enterprize Information Systems, pp. 236–243 (2004)
7. Lacher, M.S., Groh, G.: Facilitating the Exchange of Explicit Knowledge through Ontology Mappings. In: Proceedings of the 14th Int. FLAIRS Conference, pp. 305–309 (2001)
8. Qu, Y., Hu, W., Cheng, G.: Constructing virtual documents for ontology matching. In: Proceedings of the 15th International Conference on World Wide Web (WWW 2006), pp. 23–31 (2006)

9.  Doan, A., Madhavan, J., Dhamankar, R., Domingos, P., Halevy, A.Y.: Learning to Match Ontologies on the Semantic Web. Proceedings of the VLDB Journal 12(4), 303–319 (2003)
10. James, C.A., Weininger, D., Delany, J.: Daylight theory manual daylight version 4.82. Daylight Chemical Information Systems (2003)
11. Jiang, H., Wang, H., Yu, P.S., Zhou, S.: GString: A Novel Approach for Efficient Search in Graph Databases. In: Proceedings of IEEE 23rd International Conference on Data Engineering, ICDE, pp. 566–575 (2007)
12. Shasha, D., Wang, J.T., Giugno, R.: Algorithmics and applications of tree and graph searching. In: Proceedings of Symposium on Principle of Database Systems, PODS, pp. 39–52 (2002)
13. Yan, X., Yu, P.S., Han, J.: Substructure similarity search in graph databases. In: Proccedings of International Conference on Management of Data-SIGMOD, pp. 766–777 (2005)

# A Semantic Web Information Integration System Based on Ontology-Mapping and Multilevel Query Interface

Zhicheng Wan, Yifan Zhai, and Haibo Yu

Shanghai Jiao Tong University, Shanghai 200240, P.R. China
{wanzc12345,marktheone,haibo_yu}@sjtu.edu.cn

**Abstract.** The Semantic Web has introduced various distributed and dynamically increasing knowledge bases that are built on formal logic. At present, however, each knowledge base has merely a certain or some limited types of information, which are insufficient in consideration that users demands may involve extra information in most cases. Moreover, current knowledge bases are generally constructed on a mixture of ontologies, which more or less cause difficulties in the searching and integration of semantic data. Besides, a great majority of users are incapable of utilizing semantic searching. Above all, therefore, this study designed and implemented a system that allows different groups of users to query semantic information from multiple sources through a unified user-friendly interface by introducing a multilevel searching interface, which provides four fundamental input methods that allows users to generate formal query as well as proposing a processing method that achieves ontology mapping, automatic searching and integration of data from different knowledge bases. The evaluation experiments were conducted and the results of which showed that the usability and performance of our system are reasonable.

**Keywords:** information integration, ontology mapping, semantic search interface, query method.

## 1 Introduction

Semantic web technology describes accurate information with formal language which allows information to be understood and automatically processed by machines, and makes automatic and accurate information query processing possible. After years of development, it continues to grow mature, with more and more published semantic information being collected by many knowledge bases, from which one can carry out a variety of knowledge searching. On one hand, knowledge bases compose the source of network semantic information, but in most cases are just designed for a certain or some limited types of knowledge, which is not the case in real world where it involves a wide range of knowledge. At present the amount of Semantic Web data and knowledge base is growing, but none of them can cover all relevant information. Thus, a single knowledge base

G. Qi et al. (Eds.): CSWS 2013, CCIS 406, pp. 76–89, 2013.
© Springer-Verlag Berlin Heidelberg 2013

may not be able to meet users needs in querying semantic information. On the other hand, data for machine processing requires applying machine query language, which in terms of human interface is not friendly indeed. Some of the existing semantic information bases do provide query interfaces, but none of them is friendly enough. Besides, styles of different bases are distinct from one another, causing lots of inconvenience to users. The purpose of this study is to design and implement a system that allows different groups of users to query semantic information from multiple sources through a unified user-friendly interface and an automatic data integration to improve the coverage and accuracy of users information query, and to finally reach the goal of improving user's productivity.

Ordinary users appear to have difficulties even with the simplest Boolean expressions, let alone the use of the description logic formalism beneath the Semantic Web. So how could we fill the gap between the logic-based Semantic Web and ordinary users? The same question exists even among people who are acquainted with semantic web. What can be done for satisfying the needs of different target user groups? Research regarding this problem mostly focus on translation from Natural Language to query sentence and on Form-based query. Some teams[3,4] have developed auto-complete entering methods or have translated the sentence into query. Although auto-complete is easy to use, its range is narrow. The other way may cover a wider range, but it is not so accurate. As for the form based query method, though it has a fair usability, it is not universal enough to cover every occasion. Our team dealt with these problems by providing an intelligent multilevel user-friendly search interface for the Semantic Web. It relies on four basic methods for query, namely, Natural Language-based query method, Keyword-based query method, Form-based query method, Formal Language-based query method.

At the same time, the particular one single data source cannot meet the increasing needs of people who want to know more complete information in less searching times. The current diverse semantic knowledge bases do not accomplish this task because each one of them has its own ontology which sets obstacles for data retrieval. Therefore, integration is necessary. Taking this into account, we promoted some rewriting and ontology mapping work after generating the formal language query. Ontology mapping has received much attention, and many researches are working on it[14,15]. We introduced ontology mapping-based automatic information integration method to generate the mapping rules in this research. After returning the result, we merged them according to the same rules.

In section 2, we will introduce the design principle of the whole system, while in section 3, by presenting a lot of illustrations, we give implementation details of multi-level semantic search interface. In section 4, the details about how to integrate semantic Web data automatically will be explained. Section 5 shows a thorough evaluation within a number of users. The paper ends with a discussion of the limitations and vision of our approach in section 6.

## 2   System Architecture

The system was divided into two parts. One is the user interface module, and the other is the information transformation and integration module. The Fig. 1 shows the structure of our system. In the user interface module, we proposed

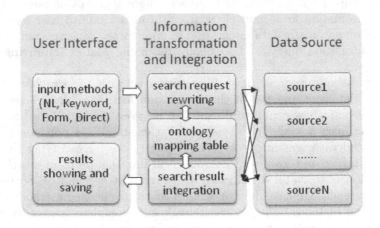

**Fig. 1.** System Structure

a multilevel query interface which takes into account of the needs of different groups of users in order to improve the usability and accuracy.

Because of different habits and proficiency in the use of Semantic Web among users, we provided four disparate query methods. The Natural Language-based method is for people who have no experience in searching on-line and want to query using daily conversation. The keyword query method is for people who have experience in using the common search engines like GOOGLE. The form based query is for those who have experience in searching semantic web. And the formal language query is for experts in this field. The following is our design in a more specific view. Firstly, we used several patterns to cover all frequently asked questions in academic field. Thus, plain English text will be processed by matching the most used query pattern. If matched, it will be transformed into the SPARQL query sentence respectively; and if not, it would match with the most similar pattern and show message offering feedback to users. As for keyword based query, users merely have to enter several related keywords. The form method used in our research constructs every token, namely, specific part of SPARQL query sentence, by leaving those parts blank, so that users could fill in them by themselves. The direct way is available as well, so that users could enter the exact SPARQL query sentence.

In the second part, we compared several most-frequently-used software to construct ontology mapping table in order to transform the ontology-local-based query into ones that based on different specific ontology. Then we rendered the transformed query to retrieve information from a number of knowledge bases in specific field, including ontology-source1, ontology-source2 etc. After getting the distributed results, we reversely merged them according to the same rules of separating. Then the final result came into being.

# 3   Multilevel Semantic Query Interface

To build a fast prototype, we focused on the academic field and used the SWRC ontology(version 0.7) as our abstract ontology. In reality, the DBLP, CITESEER and DBPEDIA work as our data source. Having these, we developed four different input methods to provide a user-friendly multi-level semantic search interface. The four methods are Natural Language-based, Keyword-based, Form-based and Formal Language-based query method.

## 3.1   Natural Language-Based Query Method

The current research in NLP(Natural Language Processing) contains a wide range of topics. The results of the evaluation show that GAPP[8] provides an intuitive and convenient way for anatomists to browse the FMA knowledge base, but its accuracy may be inferior, in that it does not have similar-match process as we do. The START question answer system[9] allows NL annotations in describing the content of a knowledge base. It has the same ambiguous problem as GAPP does; PowerAqua[1] implements similar functions, but it generates logical queries according to each ontology directly without ontology-mapping, which is lack of universality; meanwhile, both of them just use NL method to generate the query.

Because of the ambiguity in plain English, each word in the question should be confined to an exact field and a professional term in that field. The parsing has two methods to process-forward and backward mapping. Here we chose more natural one-former. As for the translation, since it is not practical to build every question-query match pairs in all fields, the field constraints are also necessary.

To get a fast and practical prototype, we focused on the academy field of Semantic Web itself. An abstract knowledge base of the field needs to be built before analysis. Considering that information people concerned are limited, we created 15 frequently asked question patterns as a start.

**Constructing Context-Free Grammar.** As stated above, according to the Internet search history statistics, we listed several frequently asked questions in academy field as following(excerpt):

1. WHAT ARE (THE) PUBLICATIONS | ARTICLES | THESES OF < AUTHORNAME >
2. WHAT ARE < AUTHORNAME >'S PUBLICATIONS | ARTICLES | THESES

3. WHAT | WHICH ORGANIZATION | UNIVERSITY IS < PERSONNAME > IN
4. ...

In the questions above, the content between the sharp braces is the subject or object's name entered by users. The specific words used in the questions are terms which are made according to the class in the SWRC ontology. The mapping relationship shows as following(excerpt):

- PUBLICATIONS | ARTICLES | THESES ↔ PUBLICATION
- ORGANIZATION | UNIVERSITY ↔ ORGANIZATION
- AUTHOR | WRITER | CREATOR ↔ AUTHOR
- ...

Based on those questions, we generated the context-free grammar accordingly, which is shown as following(excerpt):

- WHAT → IS | ARE | WAS | ORGANIZATION | UNIVERSITY | PROJECT | PUBLICATIONS | PRODUCTS
- WHERE → WAS
- WHO → IS | ARE
- ...

**Matching Sentence with Pattern.** After constructing context-free grammar, we developed an algorithm to match the entered sentences with those patterns to realize the translation work. With the grammar above, we used the forwarding match algorithms to match sentences with each pattern.

Firstly, it gets the input sentence and deletes irrelevant token such as *the*, ? etc. Secondly, it adds the ending symbol $ to the bottom of trimmed sentences. Thirdly, it starts to takes the first token in and compares this token with each pattern's current token. If the current token in each pattern is not comparable to the token taken in, this pattern is moved out from the pattern group. If matched, then remove this token in pattern. When each pattern has been matched, move on to the next token. If current token is the ending symbol, namely $, then it reaches to the end.

At last, according to every matching result, we generated the SPARQL query as Table 1 shows(taking the first pattern as an example).

**Calculating Similarity.** If the user entered sentence did not match the patterns successfully, which in most researches has been treated as an exception, we would reevaluate the similarity of each pattern with the text and show notification of ranking for users to choose from. The similarity algorithm is designed on the basis of three important perspectives: structure, token and content. Three standard indexes measuring those features are steps of our matching algorithm, coverage rate of predefined token and coverage rate of user entered word compared with search history. Taking the weight of each feature and test results

**Table 1.** Matched Query Example

| PATTERN | QUESTION | SPARQL |
|---|---|---|
| WHAT ARE PUBLI-CATIONS OF < AU-THORNAME > | What are publications of David Balcom? | SELECT distinct ?title WHERE ?author swrc:firstName David ?author swrc:lastName Balcom ?publication swrc:creator ?author ?publication swrc:title ?title LIMIT 10 |

(see section 5) into consideration, we finally set three parts proportions in similarity as 3:4:3. The first part is the degree the extent for structure matching, namely, the amount of the first match token, denoted as $D_{structure}$. The second is the coverage rate of token, denoted as $C_{token}$. The third is the coverage rate of the history user content among each pattern, denoted as $C_{content}$. The calculation formulas are shown below(for each pattern):

$$D_{structure} = \frac{N_{steps\,of\,forward-matching}}{N_{steps\,of\,current\,pattern}} \tag{1}$$

$$C_{token} = \frac{N_{matched\,tokens}}{N_{tokens\,of\,current\,pattern}} \tag{2}$$

$$C_{content} = \frac{\max(N^i_{matched\,words})}{N_{words\,of\,most\,matched\,content}}, i \in N_{contents\,of\,current\,pattern} \tag{3}$$

$$P_{similarity} = \alpha \times D_{structure} + \beta \times C_{token} + \gamma \times C_{content}, \alpha = 0.3, \beta = 0.4, \gamma = 0.3 \tag{4}$$

For example, given that the user entered sentence is:"what are books Tim have written?" and a particular pattern is the first one:"WHAT ARE (THE) PUBLICATIONS | ARTICLES | THESES OF < AUTHORNAME >". The step of matching is 2, and the total step is 5. So the $D_{structure}$ is 0.4. The matched token is "WHAT", "ARE", < NAME >, and the total amount of token is 5. So the $C_{token}$ is 0.6. If the history data has "Tim", then $C_{content}$ is 1 and the total similarity is $0.4 \times 0.3 + 0.6 \times 0.4 + 0.3 \times 1 = 0.66$. If not, the similarity depends on what content it has.

Afterwards, it sorts every pattern by their similarity, and show the top 5 most-similar patterns. Users could choose one if what they originally meant to search is among them. If not, we provide an extra add-up option. The add-up operation is to enter the custom pattern and the corresponding formal language query sentence.

## 3.2 Keyword-Based Query Method

The keywords based query method is similar to search method of common search engine like Google, Yahoo and so on. It takes in some related but single words and then speculates the exact information that users want to search. But things are a little different in semantic search. It does not retrieve the content from the website, namely, the HTML files. Instead, it mainly searches the RDF files on-line. Hermes[4] tries to translate the given keyword query into the structural query and then use a RDF triple store to return the answers. Since we did not have the condition to do that job here, we had to utilize on-line engines to accomplish this work. The current using sites are listed here:

- http://citeseer.rkbexplorer.com/browse
- http://dblp.rkbexplorer.com/browse
- http://dbpedia.org/fct/

However, instead of sending all words together, we conducted some preparation to improve the search performance before that.

The keyword based query method is related to the NL method, because of the innate similarity. So at the beginning we analyzed entered words by similar technique. At first, every word typed in would be compared with the tokens predefined in the NL patterns(like publication, writer and so on). If matched, this word would be filtered because it is useless to our search work. Nonetheless, before sending remaining meaningful words directly to the search engine, there is still some optimization work to do.

The search engine of ordinary academy website(like CITESEER) just provides the basic search function, which simply compares every tuple's entity with the whole entered words, a process users usually do not mean to go through. The accuracy and possibility of success would dramatically decrease when encountering complex occasions. Therefore, we roundly combined every possible group of these words and sent them to each available site, and finally merged all returned results. For instance, if a user enters *Tim publications*, the method would generate 3 combinations to execute-*Tim*, *publications*, and *Tim publications*.

### 3.3    Form-Based Query Method

Form based input method is originated from relational databases. It gets every part of target query from value-bind HTML components to construct the query sentence. Its usability is between Natural Language-based and Formal Language-based query method. With this method, it is easier to construct formal and accurate query without much burden on ordinary users as well. Its logically a more formal version of structural query language. Therefore we should analyze its structure at first. The well-known semantic knowledge base called YAGO has this kind of query method. But it is not as complete to include every token of SPARQL query as ours. And it lacks usability for its fixed number of WHERE conditions.

Here we still take SPARQL as an example. SPARQL query sentence has been divided mainly into four categories: SELECT, CONSTRUCT, ASK, and DESCRIBE. SELECT query is used to extract raw values from a SPARQL endpoint, the results of which are returned in a table format. CONSTRUCT query is used to extract information from the SPARQL endpoint and to transform the results into valid RDF. Each of these query forms takes a WHERE block to restrict the query, except that in the case of the DESCRIBE query, the WHERE is optional.

For the goal of searching information in Semantic Web, we mainly concerned about SELECT and CONSTRUCT query forms. As illustrated above, we transformed every part of SPARQL query sentence into blanks that users could fill in. The blank tokens include following:

- BASE "user input"
- PREFIX "user input"
- ...

Each *userinput* part was replaced by blanks that users could fill in. Meanwhile, in the WHERE component, there would be up to ten requirements given to be filled. The subjects and objects of each requirement were left as blank while the property area was generated automatically by the given ontology which, in this occasion, was SWRC. These properties were provided for choosing directly.

Apart from the tokens above, others were available for direct input. But the subject and object in the WHERE field depend on two occasions. If the content entered by users started with the character ?, it would be treated as a variable and would be left in the SPARQL query sentence. If not, it became a value which would be surrounded by apostrophe. In that way, users achieved the goal of constructing query in a self-custom way.

## 3.4 Formal Language-Based Query Method

The formal language query method is the direct and simplest method. It is for people who have expertise in the field of Semantic Web, so that they can construct query by themselves. We provided a text area for users to enter the query and used a formal language(here is SPARQL) query engine to validate and execute it or return errors.

## 3.5 Generalization

As far as we have reached, all methods discussed above involve just a single domain-*Semantic Academy* or even smaller. So how could we generalize this technique to apply it to a wider range? Although we have not commenced any actual generalization work yet, several thoughts have been brought up already.

**Generating Domain-Specific Fundamental Query Grammar.** Conducting a thorough survey to obtain the most popular questions in each field is the first step, as what we had done before, in order to build the basic layer of the whole base.

**Manually Increasing Diversity.** Granting all users the right to add non-existent fields, non-existent patterns and unorganized ontology is necessary, which made a central server or cluster indispensable. When being put in public and given certain stimuli, the intelligence of the interface would naturally grow day by day.

**Self-correction Mechanism.** The accuracy of the similarity calculating algorithm will be improved as the running times accumulate, in that it consists of history data comparison.

# 4    Automatic Semantic Web Information Integration

In this research, we conducted our own mapping approach after comparing several related products. COMA 3.0[14], S-Match and AROMA[15] could produce quite good results, but they are of no use for rewriting process or too heavy for this work to perform high efficiency for our research, which including designing mapping table, rewriting, execution and integration.

## 4.1    Creating Ontology-Mapping Table

Before starting to rewrite the query to fit real ontology, we had to build the ontology mapping table to smooth the transformation work. With some tests and comparisons[14, 15], we decided to build a ontology mapping module by ourselves, in consideration that although some tools did represent a good result, they were somehow unsuitable for our system(for reasons like they were too heavy or could not seamlessly work with our system. Here is the process we used to create the table, taking exact advantages of query rewriting:

1. Reading in all properties during the query rewriting process for future usage.
2. Finding those that have the same names in the target ontology, since they are most likely to be the correct mapping word.
3. Filtering step. Usually ontologies contain such kinds of properties as $is-About$ or $has-supervisor$ while other ones may simply use $about$ or $super-visor$; therefore we had to delete those meaningless characters to create a filter keyword list included $is$, $has$ etc. After that, repeated step (1).
4. This is a complicated process, in which we utilized a dictionary-$WordNet$[13] as linguistic knowledge, as some properties may be synonyms. For example, given name is related to $firstname$ and $forename$.
5. The special process which involves manual operation, for instance, $name$ can be divided into $firstname$ and $last\,name$, so this process has to be done manually. Fortunately, it did not take much work, since we did not have many words left in this step to process, and that most remaining words actually did not have any mapping part in the target ontology.

## 4.2   Rewriting Search Request

To generalize the query, we took the SWRC ontology again. As a matter of fact, almost none of the Semantic Web is presently using this ontology. So we had to build the mapping relationship between SWRC and the real ontology. This section describes some types of query rewriting mechanism and a basic file structure for the rewriting procedure. Here are techniques we created($S_i$ and $T_j$ denote $i_{th}$ and $j_{th}$ property in source and target ontology respectively):

**Table 2.** Rewriting Mechanism

| Name | Syntax | Description | Operation |
|------|--------|-------------|-----------|
| Replace | Si = Tj | Find one mapping word in the target ontology | Replace the prefix and property word |
| Combine | Si < Tj | Several words In the source ontology to one word in the target ontology | Combine the tuple lines in order |
| Split | Si > Tj | Find several mapping words in the target ontology | Split the tuple line in order to some tuple lines |
| Ignore | Si ≠ Tj | Cannot find corresponding mapping words in the target ontology | Ignore the whole tuple line |

**Replace.** Replace is the simplest type, which takes a prefix and property name replacement, such as $swrc : author$ to $akt : has - author$, $?publication\,source : author\,?author$ to $?publication\,target : has - author\,?author$.

**Split.** Split one sentence, for example, there is a property $source : name$, but target ontology has no full-name property, instead, it takes $lastname$ and $firstname$. E.g. $?author\,source : lastname\,David\,?author\,source : firstname\,Balcom$ can be changed into $?author\,target : name\,David\,Balcom$.

**Combine.** Combine, the opposite of previous type, for example, combine the $lastname$ and $firstname$ lines to one line.

**Ignore.** Ignore. Sometimes, unfortunately, no mapping word can be found in the target ontology, then that line will be ignored.

To execute the rewriting, we need a file structure which the program will read and look for the property name of source ontology. Fig. 2 shows a sample.

In Fig. 2, square bracket is the property name; the left part is the name in source ontology; and the right part is its corresponding property name in target ontology. The quotation mark shows how to separate the value. Besides, the order in front of it and behind it is also important as rewriting will generate

Generate Time: $GENERATE_TIME
Source: $SOURCE_ONTOLOGY
Target: $TARGET_ONTOLOGY
[title] ←→ [ has-title ]
[name] ←→ [ firstname ]" "[ lastname ]
[date] ←→ [ null ]

**Fig. 2.** Query Rewriting File Structure Sample

according to this order. If it gets null, that line will be ignored. Table 3 is an
example:

**Table 3.** Matching Result Example

| SWRC | AKT |
|------|-----|
| SELECT distinct * WHERE<br>?author swrc:firstName "David" .<br>?author swrc:lastName "Balcom" .<br>?publication swrc:author ?author.<br>?publication swrc:title ?title .<br>LIMIT 10 | SELECT distinct * WHERE<br>?author akt:full-name "David Balcom" .<br>?publication akt:has-author ?author .<br>?publication akt:has-title ?title .<br>LIMIT 10 |

### 4.3   Execution and Result Integration

After rewriting, each base-specific query will be sent to according base to execute. When every one of the specific query succeeded to return the result, we
still had to merge all valid results to finish the whole process. The Integration
work is the reverse process of ontology mapping.

After getting tuples from all formal language query engines, we checked the different tuples meta data and traced their father properties in SWRC. If matched,
these tuples would be merged into the final result. If not, they would be abandoned.
Thus we finished the whole job of cross-domain information retrieval.

## 5   Experiments and Evaluation

To evaluate the design and implementation of this system, we invited 10 persons
with different level at knowledge of semantic web to randomly give the information they want to search in academic field. Each person made 5 queries. Then
we let them use this system to search what they wanted and recorded the time
of getting the result and the rating(satisfying degree ranging from 0% to 100%)
of it, and in this way, assessed the usability, efficiency and accuracy of these
methods. The following diagrams and analysis are based on the results we got.

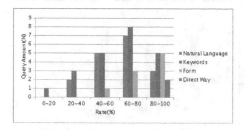

**Fig. 3.** Rates Bar Graph

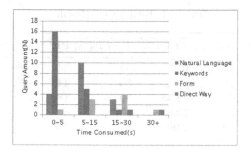

**Fig. 4.** Time Consuming Bar Graph

Fig. 3 shows the distribution of rates for each method. From Fig. 3 we find that the rates of NL and KW(Key Words) methods are more diverse. F(Form) and DW(Direct Way, implicating Formal Language-based method) are more concentrated. And the rates values of the later methods are also higher than those of the former two. These characteristics may be attributed to the accuracy of these methods themselves. However, because that the total amount of queries using first two methods is much more than the other two, it is hard to tell that the NL and KW are definitely better than F and DW. The Fig. 4 is the time consuming bar graph. From Fig. 4 we can see that the order of time consuming among 4 methods is KW < NL < F < DW. The reason for this may be ascribed to the length and complexity of input content.

Fig. 5 is the result distribution diagram. From Fig. 5, we reached the conclusion that different knowledge base would get different result when responding to the same SWRC ontology-based query. This might due to the trans-formation algorithms inaccuracy as well as the amount and type of data which they stored. Also, we found that the total number of the results was much smaller than algorithm sum of the three bases. That indicated there was a large part of data shared by all sources, which also make sense.

The Table 4 reveals the total interface result. From Table 4 we could see that more time-consuming resulted in more accuracy. The total average rate was 87 and the total average time was 8.96. These data showed that our design and implementation are strongly capable of handling different circumstances efficiently.

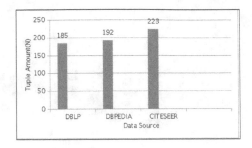

**Fig. 5.** Result Distribution Bar Graph

**Table 4.** Evaluation Result

| METHOD | $N_{queries}$ | $TIME(s)$ | $RATE(\%)$ | PROPORTION(%) |
|---|---|---|---|---|
| Natural | 17 | 9.45 | 82 | 34 |
| Keyword | 22 | 4.34 | 85 | 44 |
| Form | 9 | 12.35 | 90 | 18 |
| Direct | 2 | 30.65 | 93 | 4 |
| Total | 50 | 8.96 | 87 | 100 |

## 6　Conclusions and Future Work

In this research, we designed and implemented a system that allowed different groups of users to query semantic information from multiple sources through a unified user-friendly interface and an automatic data integration to improve the coverage and accuracy of users information query. At first, we discussed several methods to construct the query interface to search the semantic web, including Natural Language-based, Form-based, Keyword-based, and Formal Language-based query methods. And then, we constructed the means of matching SWRC ontology to specific ones to integrate all information within one interface. In order to reach the goal of high usability and accuracy, we designed different methods by using some techniques in Natural Language Processing and structured query language. Compared with similar works that have been done before, our implementation possesses higher usability within a wider range of users as well as displays its strong accuracy and efficiency in generating query. In the future, we will enable users to input domain ontology and support scalable extension for relevant domain in order to satisfy different users search demands. We could also do further research in how to improve the algorithms' accuracy and in introducing more methods to satisfy increasing needs of new perspectives.

## References

[1]　Lopez, V., Motta, E., Uren, V.S.: PowerAqua: Fishing the Semantic Web. In: Sure, Y., Domingue, J. (eds.) ESWC 2006. LNCS, vol. 4011, pp. 393–410. Springer, Heidelberg (2006)

[2]   Bowen, P.L., Chang, C.J.A., Rohde, F.H.: Non-Length Based Query Challenges: An Initial Taxonomy. In: WITS 2004, Washington, D.C., pp. 74–79 (2004)

[3]   Schwitter, R., Tilbrook, M.: Dynamic Semantics at Work. In: International Workshop on Logic and Engineering of Natural Language Semantics, Kanazawa, Japan, pp. 49–60 (2004)

[4]   Hermes: Data Web search on a pay-as-you-go integration infrastructure. J. Web Sem. 7(3), 189–203 (2009)

[5]   Andreason, T.: An Approach to Knowledge-based Query Evaluation. Fuzzy Sets and Systems 140(1), 75–91 (2003)

[6]   Minock, M.: A Phrasal Approach to Natural Language Interfaces over Databases. Umeå Techreport UMINF-05.09. University of Umeå, Sweden (2005)

[7]   Wang, H., Zhang, K., Liu, Q., Tran, T., Yu, Y.: Q2Semantic: A lightweight keyword interface to semantic search. In: Bechhofer, S., Hauswirth, M., Hoffmann, J., Koubarakis, M. (eds.) ESWC 2008. LNCS, vol. 5021, pp. 584–598. Springer, Heidelberg (2008)

[8]   Distelhorst, G., et al.: A Prototype Natural Language Interface to a Large Complex Knowledge Base, the Foundational Model of Anatomy. In: American Medical Informatics Association Annual Fall Symposium, Philadelphia, PA, pp. 200–204 (2003)

[9]   Katz, B., et al.: Natural Language Annotations for the Semantic Web. In: Meersman, R., Tari, Z. (eds.) CoopIS/DOA/ODBASE 2002. LNCS, vol. 2519, pp. 1317–1331. Springer, Heidelberg (2002)

[10]  Dittenbach, M., Merkl, D., Berger, H.: A Natural Language Query Interface for Tourism Information. In: ENTER 2003, Helsinki, Finland, pp. 152–162 (2003)

[11]  Fazzinga, B., Gianforme, G., Gottlob, G., Lukasiewicz, T.: Semantic Web search based on ontological conjunctive queries. J. Web Sem. 9(4), 453–473 (2011)

[12]  Search Interface based on Domain Knowledge Extraction: Drexel University College of Info Science & Technology Philadelphia, PA 19104, USA (2012)

[13]  Kaufmann, E., Bernstein, A.: Evaluating the usability of natural language query languages and interfaces to Semantic Web knowledge bases. J. Web Sem. 8(4), 377–393 (2010)

[14]  Aumueller, D., Do, H.H., Massmann, S., Rahm, E.: Schema and ontology matching with COMA++. In: Ã-zcan, F. (ed.) SIGMOD Conference, pp. 906–908. ACM (2005)

[15]  David, J., Guillet, F., Briand, H.: Matching directories and OWL ontologies with AROMA. ACM (2006)

[16]  Deutsch, A., Ludascher, B., Nash, A.: Rewriting queries using views with access patterns under integrity constraints. Theoretical Computer Science 371(3), 200–226 (2007)

[17]  Calì, A., Calvanese, D., Giacomo, G.D., Lenzerini, M.: Data Integration under Integrity Constraints. In: CAiSE (2006)

[18]  Fazzinga, B., Lukasiewicz, T.: Semantic search on the Web. Semantic Web 1(1-2), 89–96 (2010)

# Promoting Integrated Social and Medical Care through Semantic Integration and Context Visualization

Weijia Shen, Guotong Xie, Kavitha Srinivas, Anastasios Kementsietsidis,
Jason Ellis, Thomas Erickson, Kevin McAuliffe, Gang Hu, and Wen Sun

IBM Research Lab
{wjshen,xieguot}@cn.ibm.com,
{ksrinivs,akement,jasone,snowfall,kpmac}@us.ibm.com,
{hugang,sunwenbj}@cn.ibm.com

**Abstract.** With many disparate information systems distributed among social and medical care facilities, achieving an integrated social and medical view of people is a huge challenge. We propose a system based on semantic integration that addresses this challenge. It achieves light-weight data integration and navigation via a three-layer architecture: a virtual RDF view layer, a distributed query processing layer and a unified context view layer. This integrates information from disparate systems without cloning data, and also supports data exploration through a novel visualization.

## 1    Introduction

Social and medical care data is inherently complex because numerous organizations are involved in providing care (e.g., hospitals, nursing facilities, welfare agencies). An individual's social and medical records may be scattered across many facilities, and may be accessed and delivered using a variety of incompatible methods. As a result, the data stored at the various facilities may not be available at the point-of-care. Thus, data integration is an important aspect of a successful integrated care solution.

Semantic Web [1] technology is a useful way of addressing this issue: its benefits include integration of heterogeneous data using explicit semantics and rich explicit models for data representation and search [2]. Much work has focused on semantically formalizing and integrating different data sources [3]. For data source formalization, tools like D2R [4] enable dumping relational DB data as RDF (Resource Description Framework) data, and allow SPARQL queries on a relational DB via its SPARQL endpoint. For integrating heterogeneous data sources, different sources can be mapped to a common ontology (e.g. [5]). In addition, federated query processing is often used since multiple SPARQL endpoints are involved [6].

A challenge for federated SPARQL query processing is to design an architecture that does not rely on one centralized processing unit but rather uses a peer-to-peer distributed approach to facilitate adding and removing data sources. Another challenge is that once the data has been integrated, it must be presented in a way that a wide range of users in different organizations and facilities can understand and use.

G. Qi et al. (Eds.): CSWS 2013, CCIS 406, pp. 90–97, 2013.

In this paper we propose a light-weight data integration system that addresses these challenges by using a three layer architecture: a virtual RDF view layer, a distributed query processing layer, and a unified context view layer.

## 2    System Design

The three layer architecture (Fig. 1) consists of a virtual RDF view layer, a distributed query processing layer, and a unified context view layer (Fig. 1), where the reference ontology is the semantic model for the virtual RDF and query processing layers.

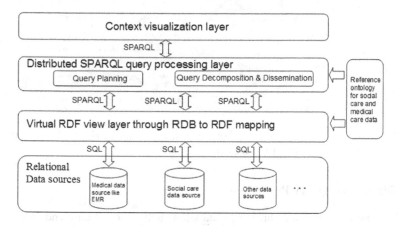

**Fig. 1.** Architecture overview

### 2.1    Virtual RDF View Layer

This layer is for semantically formalizing different data sources. A platform called Semantic Data Access (SeDA) platform (an in-house product of IBM) is adopted – it provides a scalable and efficient semantic query interface for traditional relational databases, just like D2R [4] and uses the same mapping language called D2RQ Mapping Language as D2R. It enables relational data to be exposed as virtual RDF graphs and queried using SPARQL queries. Mappings based on a unified reference ontology are defined between the RDF virtual graph and different database schemas. These mappings enable the SeDA engine to translate SPARQL queries into to SQL statements. Fig. 2 shows an example of how the mapping file (text on the right) and the unified reference ontology (on the left) are related. The reference ontology on the left has not only classes related to social care like "Facility" and "Marriage", but also includes clinical related class like "LabTest" and "Medication". For "Patient" class, it has many properties, where one of them is called "hasRisk", this property is to be mapped to a specific database table column.

To build a unified reference ontology (semantic model) for integrated social and medical care, we use the HL7 reference information model [7] to model clinical elements and a social care taxonomy [8]. We also reuse some classes and relations from ontologies like DBpedia, FOAF, Dublin Core, etc. Note that we only intend to map a small but relevant subset of data to the reference ontology; this greatly reduces the burden of data mapping and ontology building. With this layer, different databases could be accessed by issuing SPARQL queries with a unified reference ontology.

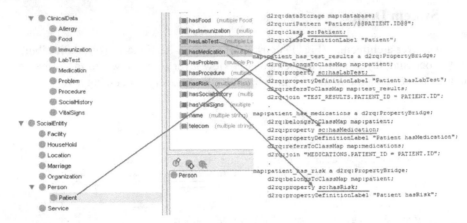

**Fig. 2.** An example showing the relations between a mapping file (underlined text on the right) and the unified reference ontology (on the left)

## 2.2    Distributed Query Processing

One of the issues common in bridging data silos such as social care and medical care is the issue of data ownership. That is, because data owners and access control layers differ, sharing data across silos via a single data warehouse is not an option – a distributed querying architecture is required. Even in a federated architecture, one needs to maintain a loose coupling between the data silos. Our approach is to map only the relevant subsets of each silo into a single unified semantic model. This model is declarative and in OWL, so it serves as a layer of abstraction that insulates the unified model from changes in the schema of each individual data model. Also, by ensuring that only a small, relevant subset of the data are mapped to the semantic model, the cost of mapping across the schemas for integration is reduced, and can be done on an incremental basis, since the mappings themselves can be added incrementally. We believe the low barrier to entry provided by this integration model is a key characteristic that makes it attractive for many integration tasks.

Our mechanism for loose coupling across silos is to use a peer-to-peer distributed architecture over a centralized store: this makes it easy to add and remove sources. Overall, the distributed querying layer has a registration service that allows multiple sources to register themselves. Each new node simply adds its location and its mappings to a common unified model. To answer a SPARQL query, each data source ultimately uses mappings to the semantic model to translate it back to a SQL query.

## 2.3    The Social Context Visualization

A further challenge of integrating medical and social data is to make it understandable to the various care workers in different organizations and facilities. Here the aim is to provide a visualization of the "social context" of the individual – "the patient" – who is receiving medical and social services. By "social context" we mean the factors – people, programs, organizations and the relationships among them – that affect the patient's daily life, and particularly his or her well-being. The visualization is intended to enable caregivers to devise appropriate solutions to the patient's problems.

To develop the social context visualization we began with a type of diagram called a genogram [9], adopted from the social care domain. Genograms resemble a family tree with complex symbolic annotations, and are used to show the structure of an extended family with emphasis on emotional connections and challenges faced by family members. For example, the genogram shown in Fig. 3 shows a divorced couple (Bob and Sue) with two children, as well the couple's siblings and parents, and the emotional relationships among them. Genograms are useful because they enable the integration of a wide range of data types in a single, person-centric representation.

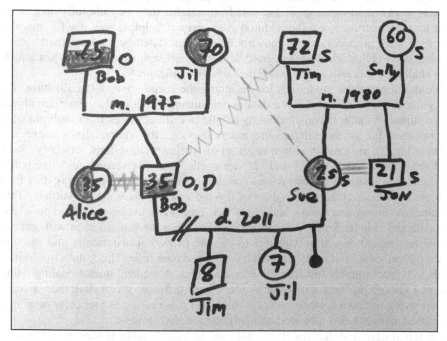

**Fig. 3.** An example of a hand-drawn genogram

But while genograms were a useful starting point, they were far from a perfect fit for our needs. Genograms began as ad hoc, temporary, hand-drawn diagrams for specialists, and needed to be re-designed to be a general purpose, persistent, computer-based visualization that could support the wide variety of tasks in the medical and social care domains. Adaptation was needed.

In thinking through the re-design of genograms, it is useful to reflect on the concept of data silos. Silos are generally considered to be problematic, and it is certainly true that their existence creates barriers to the integrated use of data by organizations. However, it is important to recognize that silos are not arbitrary – data silos reflect the way that work is organized (and thus the way that data is produced, maintained and used). That is, for any data silo, there are people acting in one or more roles who create, use and modify the data in that silo and as a consequence have some understanding of its characteristics. For example, in the case of a medical data silo, people acting in the roles of physicians, nurses, and care coordinators will use the data, and all will likely have some understanding of it (e.g., what a "blood glucose level" is and what normal values of it are). Similarly, in the case of a social work data silo, people acting in the roles of social workers, behavioral health specialists and care guides will use the data, and will likely have some understanding of it (e.g., what "SNAP" is, and its eligibility requirements). Finally, in both cases, users in various roles will have practices for working with their data (e.g., standard views) that have evolved in the context of their (siloed) work practices.

This interrelatedness of siloed data and work practices are important because when we begin to break down the silos we run the risk of disrupting work practices. When workers from one silo are exposed to data from another they may not fully understand what they see; furthermore, the addition data, even if helpful, may not fit smoothly with their work practices (e.g., it may not be accommodated by their standard views). The key point is that when we integrate data from different silos, it is not just a technical challenge, it is also a user experience design challenge.

We developed three guidelines for designing the social context visualization. The first guideline is to ensure that the data is consumable. That is, the genogram should be simplified so that it is consumable by people in different roles from multiple silos. For instance, the arcane relationship notations (e.g., the various lines connecting nodes in Fig. 3), are simplified to a small set of labeled relationships; similarly, shorthand annotations like "O", "S" and "D" are spelled out. The second guideline is that the visualization should be interactive. Another consequence of integrating data from multiple silos means that there is a lot of it – too much to show on the surface of the visualization at any one time – and so various methods must be used to expose relevant data and hide irrelevant data. In addition, because the visualization will serve as a persistent record, it will be valuable to be able to view it historically and examine how a patient's data and social context has changed over time. Third, the visualization needs to be configurable for different uses and roles. A medical doctor dealing with a patient's allergy problem will want to see a very different set of data than a social worker trying to find a foster home of a child, and those acting in particular roles may have standard views that they wish to apply to their data subsets.

An initial design sketch for the social context visualization is shown in Fig. 4. It simplifies the traditional genogram by retaining the familiar family tree structure, but uses various interaction mechanisms – hovering, filtering and pinning – to display information. Individuals are shown as squares or circles (indicating male or female); next to each shape is data about their assets (above right) and problems (lower right). More information can be accessed by clicking on the four lobes associated with each shape: the upper right lobe reveals assets; the lower right reveals challenges; the upper left shows a vulnerability index; and the lower left basic demographic data. People are connected vertically by solid lines indicating parent-child relationships, and horizontally by labeled lines indicating marital relationships; clicking on the adjacent triangles reveals the history of a relationship. The view at the left shows only the household; the view at the far right expands the scope of the visualization to include extended family, and adds a layer of emotional relationships among family members.

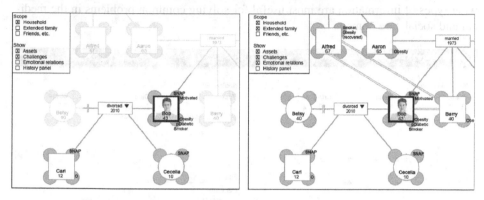

**Fig. 4.** Two views of the social context visualization design sketch

To better understand how the visualization functions, and why it would be useful, let's walk through an example of how it would be used. Let's assume that a patient named "Bob" arrives for a medical appointment. His physician, by quickly glancing at the social context visualization embedded in Bob's medical record, can quickly see that Bob is recently divorced with two children (Fig. 4, left). This helps remind him of who Bob is (after all, the physician sees hundreds of patients a year), and enables him to start out the visit by asking Bob how his kids are, and how he's faring after his divorce. Looking more closely, the physician can use the annotations to assess his patient's assets and challenges: in this case, he sees that Bob is highly motivated and is in the SNAP (Supplemental Nutrition Assistance Program) program (assets), but is a smoker, obese and pre-diabetic (challenges). He also knows that the fact Bob is eligible for SNAP means that he has financial challenges, which may add stress to Bob's life, and make certain types of remedies for his problems more difficult.

As the visit continues, the physician can expand the view to include the extended family (Fig. 4, right), and see that problems like smoking and obesity are common in Bob's family. This could support a discussion of Bob's challenges in the context of his extended family, and get him to think about some of the longer term consequences (e.g., his Uncle Aaron, who is obese, now has knee problems and difficulty getting around). The physician can also see that some members of his family – for example, his Uncle Alfred – have overcome obesity and smoking. In addition, by examining the emotional relationship layer, the physician can see that Bob and his Uncle Alfred have a close relationship (indicated by the double green lines), and so perhaps could serve as a source of mentoring and support as Bob tries to address his challenges.

While this example is fictional, it illustrates the potential utility of being able to make use of data integrated from a variety of different sources. Understanding the patient's social context enables care provides to consider a broader range of solutions.

This visualization is still under development. Initial paper sketches and mockups have been refined, and an interactive prototype is under development (Fig. 5. shows its current state). The next step is to carry out formative user testing in which users are asked to carry out common tasks in the social work and medical care domains; the testing will enable the interaction mechanisms to be refined, and assist in determining what types of information are most useful for solving common problems in the medical and social care domains.

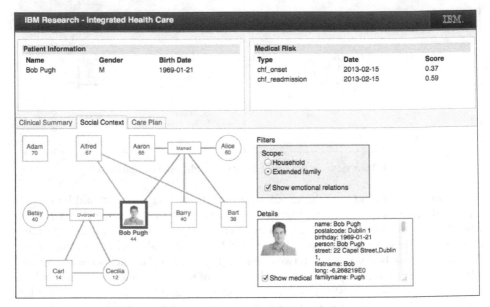

**Fig. 5.** Interactive prototype of the social context visualization

## 3    Conclusion and Future Work

This paper has described a novel system for an ongoing project that aims to achieve light-weight data integration and data exploration through semantic integration and novel context information visualization. A new context visualization approach has been discussed and presented in details. Its aim is to support the integration of social and medical care data to promote better integrated care services and context visualization. Initial testing on this prototype shows promising results. Moreover, it could serve as a general purpose data integration reference system for other domains as well. This is because the RDF view layer and query processing layer are independent of the data domains – instead they rely on a defined ontology and mappings. However, the context view layer may need to be adapted to new data domains. Therefore, extending the context view layer's ability to support more visualization patterns is our future work. Other future work includes testing the system's performance under real life scenarios and improving it according to care providers' comments.

# References

1. Berners-Lee, T., Handler, J., Lassila, O.: The Semantic Web. Scientific American (2001)
2. Baker, C.J.O., Cheung, K.H. (eds.): Semantic Web: Revolutionizing Knowledge Discovery In The Life Sciences. Springer, New York (2007)
3. Feigenbaum, L., Herman, I., Hongsermeier, T.: The Semantic Web in Action. Sci. Am. (2007)
4. Bizer, C., Cyganiak, R.: D2R Server - Publishing Relational Databases on the Semantic Web. In: 5th International Semantic Web Conference (2006)
5. Chen, H., Ding, L., Wu, Z., Yu, T., Dhanapalan, L., Chen, J.Y.: Semantic Web for Integrated Network Analysis in Biomedicine. Brief Bioinform. 10, 177–192 (2009)
6. Buil-Aranda, C., Arenas, M., Corcho, O., Polleres, A.: Federating Queries in SPARQL1.1:Syntax, Semantics and Evaluation. Web Semantics: Science, Services and Agents on the World Wide Web (2013)
7. HL7 Reference Information Model,
   http://www.hl7.org/implement/standards/rim.cfm
8. Social Care Online: taxonomy,
   http://www.scie.org.uk/publications/misc/taxonomy.asp
9. McGoldrick, M., Gerson, R., Petry, S.: Genograms: Assessment and Intervention, 3rd edn. (2008)

# A Subject Partitioning Based SPARQL Query Engine and Its NoSQL Implementation

Chuanlei Ni, Wei Hu, and Yuzhong Qu

State Key Laboratory for Novel Software Technology, Nanjing University, China
nichuanlei@gmail.com, {whu,yzqu}@nju.edu.cn

**Abstract.** The schema-free nature of RDF makes the processing of its query language SPARQL challenging. In this paper, we propose a subject partitioning approach to storing RDF data using the unit of entity document to reduce join operations. To transform a query graph to a query plan, we design a size estimation model specific to SPARQL processing. We implement a prototype system called sp-NoSQL over the MongoDB NoSQL database. By comparing with two representative SPARQL query engines, we evaluate the effectiveness of our approach.

## 1 Introduction

The Semantic Web is an important trend for the next generation of the World Wide Web. The Resource Description Framework (RDF) is a language for representing information about resources on the Web. In RDF, all data items are represented in the form of ($subject, predicate, object$) triples, and a set of RDF triples form a directed RDF graph.

SPARQL [1] is the standard language for querying RDF data. SPARQL to RDF is just like SQL to relational databases. SPARQL query engine is an important infrastructure to access massive RDF data. The schema-free nature of RDF makes the processing of its query language SPARQL challenging.

This paper introduces a SPARQL query engine based on the idea of subject partitioning to deal with some vital challenges. Our contributions in this paper are listed as follows:

1. We propose to use subject-partition for the efficient storage of RDF data. This method can reduce the number of join operations by dealing with the Basic Graph Patterns (BGPs) of SPARQL queries;
2. We implement the above method based on document stores. We develop a prototype system named sp-NoSQL, and conduct experiments to show its advantages of dealing with SPARQL queries. To the best of our knowledge, this is the first work that combines SPARQL processing with document stores;
3. We present a specific selectivity estimation mechanism to tackle the join order problem in processing SPARQL queries with the balance between the system resource consumption and the estimation accuracy.

G. Qi et al. (Eds.): CSWS 2013, CCIS 406, pp. 98–105, 2013.

The rest of this paper is organized as follows: Section 2 introduces related work; Section 3 describes our query processing algorithm; Section 4 presents the system implementation based on document stores; Section 5 reports our experimental results and analysis; We concludes this paper with future work.

## 2   Related Work

SPARQL query engine is a necessary infrastructure to support the growth of Semantic Web, and this topic has attracted a lot of researchers. Existing SPARQL query engines can be divided into three categories:

1. Native RDF storage. The representative is RDF-3X [2], which uses a RISC-style system architecture and achieves the state-of-art query processing efficiency by massive indexes;
2. Using existing relational database as storage layer. The RDF triples are stored in three-columns tables in database. The representatives are 3store [3], Jena SDB and so on. The RDF stores belonging to this category and the native RDF stores can all be considered as three-columns based stores;
3. Also reusing existing relational database as storage layer, but the idea of property table is used to store the triples. Both Jean and Sesame have the implementations based on this idea [4,5].

For a database management system, query optimization has very important impact on the entire performance, and determining join orders is the main place at which the query optimization plays role. Thus, there are a lot of works dedicated to the optimization of the join order by collecting the appropriate statistics from the RDF data stored, for example [6].

There are works that aim at reducing the number of join operations needed in the query processing. This is the motivation of property table. The basic idea of property table is to mine the frequent patterns in the RDF data and store the RDF triples with frequent co-occurrence properties in one database table. This storage mechanism can often speed up the query processing.

However, because of the diversity and sparsity nature of RDF data, the schema of property table is hard to predict. So Sesame mines the frequent pattern using data mining techniques, and Jena requires the users specify the property table schema. But the updates to RDF data is a big challenge for Sesame; while Jena's method loses the scalability even with having put heavy burden on the users. At the same time, it is common that a property has different values for the same subject in RDF data. It is somehow complex for property table system to deal with this problem.

To reduce the number of join operations, our sp-NoSQL extends the idea of property table while without requiring the users to specify the database schema. To decide the join order, we propose a strategy to collect the statistics specific to the SPARQL query processing.

# 3   Query Processing Algorithm

## 3.1   Basic Definitions

For the convenience of presentation, we make the following conventions:

Let $\mathbf{I}$, $\mathbf{L}$, $\mathbf{B}$ be the set of all IRIs, the set of all RDF literals and the set of all blank nodes, respectively. And let $\mathbf{T}$ be $\mathbf{I} \cup \mathbf{L} \cup \mathbf{B}$.

Let $\mathbf{V}$ be an infinite set that disjoint with $\mathbf{T}$. A triple pattern is the element of $(\mathbf{T} \cup \mathbf{V}) \times (\mathbf{I} \cup \mathbf{V}) \times (\mathbf{T} \cup \mathbf{V})$. For a triple pattern $t$, we use $t.s$, $t.p$, $t.o$ to denote the subject, predicate, object of this triple pattern, respectively. The *basic graph pattern* (BGP) is a collection of triple patterns.

**Definition 1 (Subject Triple Block (STB)).** *Subject triple block is a special kind of basic graph pattern where all the triple patterns have the same subject.*

**Definition 2 (Virtual Table).** *Virtual table refers to the set of tuples of the query result of a basic graph pattern. The schema the table is the variables appearing in the basic graph pattern.*

**Definition 3 (Entity Document).** *Entity document is a collection of RDF triples where all the RDF triples have the same subject.*

In sp-NoSQL, the RDF data is loaded into the document store in the unit of entity document. To execute the SPARQL query, we get the virtual tables for the basic graph pattern. At last, the virtual tables are joined to get the final query results. The rest parts of this section will detail the algorithm of generating query plan for the basic graph pattern while the execution of the query plan will be explained in the next section.

## 3.2   BGP Partition Based on Subject

To reduce the number of join operations effectively, we propose an algorithm based on subject partitioning shown in Algorithm 1. We analyzed the query log of DBpedia, and found that about 95% queries in the log contain only one subject-triple-block and about 99% queries contain at most two subject-triple-blocks. This statistics shows a great future of our algorithm that aims at reducing the join operations and sometimes even omits the join operation.

---

**Input**: A basic graph pattern, *bgp*
**Output**: A virtual table, *vt*
1   $g \leftarrow$ partition-by-subject(*bgp*);
2   *tree* $\leftarrow$ decide-join-order($g$);
3   **return** *tree*;

---

**Algorithm 1.** Query plan generation based on subject partitioning

Firstly, we partition the basic graph pattern such that all the triples have the same subject belong to the same subject-triple-block. We treat each subject-triple-block as a node; and if two subject-triple-blocks share variable, they have an edge. The query result of every subject-triple-block is the virtual table which is a corresponding concept of base table in the relational database community.

Secondly, we get the join tree from the query graph. This is a classical problem in the relational database field [8]. However, when applying those classical algorithms to our subject partitioning algorithm, we face a basic problem: we lack the statistics about the number of tuples in virtual tables which is obvious in relational database. This case is a little different from the relational database that the traditional strategy is useless.

To solve this problem, we propose a SPARQL specific selectivity estimation method. Given a subject-triple-block $B$, we can estimate the size of corresponding virtual table as shown in Algorithm 2:

---

**Input**: Subject triple block, $B$
**Output**: Num of rows of virtual table, $N$
1 **if** *the triple patterns in* B *have constant subject* **then**
2 $\quad$ | $\quad$ **return** $N \leftarrow 1$
3 **end**
4 $pL \leftarrow \{t.p | t \in B\}$;
5 template$(B) \leftarrow \{(p, q) | p \in pL, q \in pL\}$;
6 distTemplate$(B) \leftarrow (p, q)$;
7 $\quad$ s.t. $co(p, q) \leq co(p', q')$, for $\forall (p', q') \in$ template$(B)$;
8 **if** $p$ *has constant object* **then** $dv_1 \leftarrow dv(p)$;
9 **else** $dv_1 \leftarrow 0$;
10 **if** $q$ *has constant object* **then** $dv_2 \leftarrow dv(q)$;
11 **else** $dv_2 \leftarrow 0$;
12 **if** $max(dv_1, dv_2) > 0$ **then**
13 $\quad$ | $\quad$ **return** $N \leftarrow co(p, q) / max(dv_1, dv_2)$
14 **else**
15 $\quad$ | $\quad$ **return** $N \leftarrow co(p, q)$
16 **end**

**Algorithm 2.** Size estimation for virtual table

---

$Co(p,q)$ is the number of entity documents that contain both property $p$ and property $q$. When $p$ equals to $q$, $co(p,q)$ is the number of entity documents that have the property $p$. $dv(p)$ is the number of different values for property $p$ in the whole stored dataset.

The appearing properties in the RDF dataset is dependent, and the co-occurrence has some patterns. The motivation of algorithm above is to use the co-occurrence patterns to make the size estimation more accurate. We will explain the steps of executing query plans to get virtual table in next section.

## 4   System Implementation

The system implementation contains two parts: the storage of entity documents and getting the query results through subject-triple-block since the leaves of the join tree are the subject-triple-blocks.

The NoSQL databases are popular recently since the ACID is too strong for some Web applications. Existing NoSQL databases can be classified into four categories [9]:

1. Key-value stores, such as Redis, SimpleDB;
2. Document stores, such as MongoDB, Riak, CouchDB;
3. Column-family stores, such as are Cassandra, HBase;
4. Graph databases, such as Neo4J, GraphDB and so on.

There are some works in using the NoSQL systems as the storage for RDF data to support SPARQL query language. For example, [10] deals with the issues of storing RDF data in the key-value stores. [11] makes the exploration of using HBase as the storage layer of RDF data. Also, there are a lot of works that use graph databases to store RDF data, e.g., Neo4j natively has the capability of supporting SPARQL. In this paper, we use the document store MongoDB as the storage layer for sp-NoSQL. To the best of our knowledge, this paper is the first work on using document stores for RDF data.

sp-NoSQL stores every entity document as the documents in MongoDB and uses the special key "ID" to store the URI of each entity. Except for the "ID" key, other keys are the properties describing this entity. Thus, the values of these keys have the form of collection to deal with the issues that the properties in RDF dataset may contain more than one values.

We use strings to store the URI of an entity, while we store the RDF literal as complex structure. The complex structure is a nested document that may contain three different keys: the key "value" stores the lexical form of RDF literal; the key "datatype" stores the datatype of the RDF literal; and the key "lang" stores the language tag. "datatype" and "lang" may appear at most one due to the constraint of RDF syntax.

To get the query results, we need transform the subject-triple-blocks into the queries can be understood by MongoDB. The queries for MongoDB is presented as documents too. The transformation procedure can be summarized as follows:

1. If the the subject-triple-block's subject is not variable, set the "ID" value be the entity's URI;
2. If the predicate in a triple pattern is known while object is unknown, use the predicate as key and {$exists: true} as the corresponding value;
3. When both the predicate and object in a triple pattern are known, use the predicate as the key and the object as the corresponding value.

However, our prototype system is inefficient to deal with the SPARQL queries with unknown predicates due to the keys in the MongoDB queries must be specified. This problem will be fixed in the future work.

After getting the virtual tables for the subject-triple-blocks, we just use the classical technology in relational database community to combine the virtual tables to get the final results.

## 5    Experiments

In this section, we experiment our system with the Berlin SPARQL Benchmark (BSBM) [12]. For comparison, we also evaluate the performance of Jena TDB, which is a typical native RDF store, and Virtuoso, which performs impressively on many different benchmarks.

### 5.1    Benchmark

BSBM [12] features RDF data scalable to an arbitrary size and 12 extensional queries revealing a variety of properties to simulate real use cases. In our test, we generate three datasets with 1 million, 25 million and 50 million RDF triples. We refer them as 1M, 25M and 50M datasets, respectively.

### 5.2    Systems under Test and their Configuration

The experiment was conducted on a personal computer (processor: Intel Q9500 2.83GHz; memory: 8G DDR2 667; hard disk: 500G (7,200 rpm) SATA2 running Windows 7 64-bit). Java version 1.6.0_07 was used and all three systems under test were run under the Java HotSpot(TM) 64-bit Server VM. The SUTs are all be configured to their best performance.

### 5.3    Overall Performance

The central performance metric of the BSBM benchmark are *Query Mixes per Hour* (QMpH). Because the query 11 involves unknown predicate which can not be processed by sp-NoSQL now and query 6 consumes too many system resources to slow down the performance of all the SUTs, we excluded these two queries to measure the overall performance of SUTs. Fig. 1 shows the performance of SUTs in QMpH excluding query 6 and 11.

As the results show that Virtuoso gets the best overall performance while sp-NoSQL is slightly worse, while TDB is the worst one. With the size of the RDF dataset increasing, the scalability of Virtuoso and sp-NoSQL is much better than that of Jena TDB. The overall analysis shows that the performance of subject-partition storage mechanism is comparable with Virtuoso's query processing algorithm, and has much advantage over Jena TDB. In the next subsection, we will analyze the performance of SUTs by individual query to have a deep understanding of the three systems.

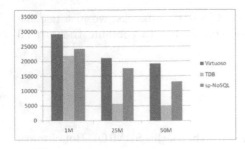
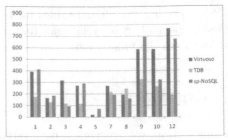

**Fig. 1.** Overall performance in QMpH, excluding query 6

**Fig. 2.** Performance of individual queries in QpS on the 50M dataset

## 5.4   Performance by Individual Query

In order to allow more detailed analysis, the benchmark results are also reported on a per query type basis. The *Queries per Second* (QpS) measures the number of queries for a specific type that were answered by the SUT within a second. Fig. 2 shows the performance of individual queries in QpS on the 50M dataset, which is the most typical situation.

From the results, we can summarize that, on some types of queries (e.g., 3, 7, 8, 10, 12), Virtuoso performs better than sp-NoSQL, while on the other types of queries (e.g., 1, 2, 4, 5, 9), the case is reversed. TDB is the slowest on almost all the queries.

Through the detailed analysis, we can reason the strengths and the weaknesses of our subject partitioning algorithm. For the queries that our prototype system performs much better than the other two, they have the same features that the join operations can be significantly reduced by subject partitioning. On the other cases, these queries may have other SPARQL features such as optional clause and union clause that we have not made optimization for. However, the analysis above is sufficient to demonstrate that sp-NoSQL deals with join operations efficiently.

## 6   Conclusion and Future Work

In this paper, we proposed a new storage model for RDF data based on subject partitioning, which can reduce the number of join operations when conducting SPARQL query processing. This method can fulfill the expected performance of property tables but avoid specifying the data schema in designing property tables. Based on a popular document store MongoDB, we implemented a prototype system named sp-NoSQL to verify our approach as compared with other two popular SPARQL query engines.

There are many unfinished issues in our current system. The unknown predicates in SPARQL queries will hurt the performance of sp-NoSQL. Also, this paper focuses on the basic graph patterns mainly, however, the optimization for the optional and union clauses is an important part of our future work.

**Acknowledgements.** This work is funded by the National Natural Science Foundation of China (Nos. 61223003 and 61003018), the National Social Science Foundation of China (No. 11AZD121), and the National Research Foundation for the Doctoral Program of Higher Education of China (No. 20100091120041).

# References

1. SPARQL Query Language for RDF. W3C Recommendation (January 15, 2008), http://www.w3.org/TR/rdf-sparql-query/
2. Neumann, T., Weikum, G.: The rdf-3x engine for scalable management of rdf data. The VLDB Journal 19, 91–113 (2010)
3. Harris, S., Gibbins, N.: 3store: Effcient bulk RDF storage. In: Proc. of PSSS 2003, pp. 1–15 (2003)
4. Abadi, D.J., Marcus, A., Madden, S.R., Hollenbach, K.: Scalable semantic web data management using vertical partitioning. In: VLDB, pp. 411–422 (2007)
5. Wilkinson, K.: Jena Property Table Implementation. In: SSWS (2006)
6. Stocker, M., Seaborne, A., Bernstein, A., et al.: SPARQL basic graph pattern optimization using selectivity estimation. In: Proc. WWW, pp. 595–604 (2008)
7. Sidirourgos, L., Goncalves, R., et al.: Column-Store Support for RDF Data Management: not all swans are white. In: Proc. VLDB, pp. 1553–1563 (2008)
8. DeHaan, D., Tompa, F.W.: Optimal top-down join enumeration. In: Proc. SIGMOD, pp. 785–796 (2007)
9. Hecht, R., Jablonski, S.: NoSQL Evaluation: A Use Case Orinted Survey. In: Proc. CSC, pp. 336–341 (2011)
10. Bugiotti, F., et al.: RDF data management in the Amazon cloud. In: EDBT/ICDT Workshops, pp. 61–72 (2012)
11. Khadilkar, V., Kantarcioglu, M., Castagna, P., et al.: Jena-HBase: A Distributed, Scalable and Efficient RDF Triple Store. In: Proc. ISWC, Posters & Demos (2012)
12. Bizer, C., Schultz, A.: The Berlin SPARQL Benchmark. IJSWIS 5(2), 1–24 (2009)

# An Analysis of Data Quality
# in DBpedia and Zhishi.me

Yanfang Ma* and Guilin Qi

School of Computer Science and Engineering, Southeast University,
Nanjing 210096, China
{myf,gqi}@seu.edu.cn

**Abstract.** Linked Data has experienced an accelerating growth since
it was launched on 2006. While an increasing amount of RDF data is
available on the web, errors also proliferate, thus the quality of Linked
Data has drawn more and more public attention. Since the quality of
data in some way affects the reliability and efficiency of web applications
consuming Linked Data, demand for quality analysis of Linked Data becomes increasingly imperative. In this paper, we present some problems
concerning the quality of Linked Data. These problems are discovered
through our analysis on two cross-domain RDF datasets: DBpedia and
Zhishi.me, both of which are based on automatic extraction of resources
from existing encyclopedias. Some of the problems can be detected simply by SPARQL queries, while others cannot. For every problem listed
in this paper, we present a method for the detection of it. Besides, we
do experiments to demonstrate the validity of our methods.

## 1  Introduction

Recent years we have witnessed the rapid development of Linked Data. Many
companies, organizations and even governments have realized the importance of
it and adopted the principles of Linked Data [2] to publish their own data. The
amount of structured RDF data is keeping growing, and at the same time, web
applications that consume Linked Data are also emerging continually. Unfortunately, since the rapid development of the semantic web has inevitable led to
many teething problems, web applications often need to face challenges coming
from inaccurate knowledge representation. Therefore, improving the quality of
published RDF data has become a key factor in prompting the development of
the semantic Web. The first step of improving the quality of data is to identify
problems in datasets. Some issues can be identified simply by SPARQL queries,
while the detection of others would not be so simple. To detect problems manually is unrealistic since what we are faced with is a vast amount of RDF triples. In
that case, it is necessary to find a way for automatic or semi-automatic detection.

DBpedia [1], as the core dataset of Linked Data Cloud, comes from the extraction of structured information from Wikipedia. The quality of DBpedia becomes
more and more important since it is broadly used in web applications. In order

---

* Corresponding author.

G. Qi et al. (Eds.): CSWS 2013, CCIS 406, pp. 106–117, 2013.
© Springer-Verlag Berlin Heidelberg 2013

to promote the development of Chinese Linked Data, Zhishi.me [7] was proposed. It is the first effort to publish large-scale Chinese Linked Data. Data in Zhishi.me origins from three Chinese encyclopedias: baidubaike, hudongbaike, and the Chinese version of Wikipedia. We should notice that, resources from all these encyclopedias mentioned above are created by various contributors manually, which inevitably leads to the error-prone feature of DBpedia and Zhishi.me.

The main contributions of this article are as follows:

- Since in Zhishi.me, different IRIs may represent the same property, we propose an approach to detect properties which have the same semantics.

- We present a method to check whether subclass-of relations are defined inversely. This method take the number of individuals of parent class and its subclass into consideration, then check the correctness of the subclass-of relation in Wikipedia.

- We have found that in DBpedia and Zhishi.me, there exist classes which share the same identifiers with their instances. To detect such classes, we give a SPARQL query and use it in DBpedia and Zhishi.me to get some statistical results.

- To detect the misuse of properties, we first discuss the misuse of skos:broader and skos:narrower. This kind of errors can be detected simply by SPARQL queries. Second, misuse of datatype property and object property has also been discovered in Zhishi.me, so we propose a method to find all the properties in Zhishi.me which are used both as datatype property and object property.

- For our own proposed detection methods, we do experiments on DBpedia and Zhishi.me to illustrate their validity.

The rest of this paper is organized as follows: Section 2 recapitulates previous works in the field of Linked Data quality analysis and error detection. In Section 3, we first present problems which have been discovered through analysis on RDF dataset, then we describe our methods for the detection of each problem. In Section 4 we report our experimental results to show the validity of methods we have proposed. Section 5 summarizes works in this paper and provide an outlook for future research.

## 2   Related Work

As data volume increases, errors are also prevalent. Thus, data quality has become a hot research topic. There has been many research work concerning the quality of data in traditional databases. In [12] data quality was defined to be the capacity of data to be used effectively, economically and rapidly. Three approaches to deal with data quality were introduced in [12]: intuitive, theoretical and empirical. Data cleaning, which aims to improve data quality by detecting and removing errors and inconsistencies from data, was discussed in [9]. [9] discussed data quality problems in single source and multiple sources, as well as on instance level and schema level. Major steps for data cleaning were also listed in this paper. However, [9] mainly discussed data cleaning problems in data warehouses rather than in Linked Data.

There have been some studies on the detection of inconsistencies and errors in Linked Data. Some tools have been designed for syntactic validation such as VRP [1]. ORE [6] was a tool for repairing and enriching OWL ontologies. It used supervised learning for extending an ontology, and provided guides for users to resolve various inconsistencies in OWL ontologies. In [4] Hogan et al. have focused on different symptoms concerning incompleteness, incoherence, hijack and inconsistency. They also proposed recommendations from the perspective of publisher and consumer respectively to avoid those problems. Besides, an online-tool RDF:ALERTS was designed to detect those errors. Later in [5] they conducted quantitative empirical analysis on a crawl of 4 million RDF/XML documents. The analysis was based on a list of fourteen concrete guidelines which were given in the "How to publish linked data on the web" tutorial.

Inconsistencies in DBpedia has been studied in many research work. Domain and range inconsistencies of DBpedia were discussed in [8] and [10]. In [8], Péron et al. searched for inconsistency patterns using SPARQL queries. The approach proposed in [10] was to transform semantic errors into logical ones. They followed methods from Inductive Logical Programming (ILP) to extend the axioms of the underlying ontology by determining new domain and range restrictions as well as class disjointness axioms. That is to say, before finding inconsistencies, they first enrich the ontology. Ontology enrichment is a necessity in the detection of some logical inconsistencies. This topic has also been discussed in [11] and [3]. [11] focused on finding axioms that were captured by OWL 2 EL using association rule mining, while [3] only focused on learning class disjointness, and several inductive approaches were presented in it.

Although there has been some work on detecting errors in DBpedia, there is no work on analyzing errors in Zhishi.me. Compared with DBpedia, the main difficulty for the analysis of Zhishi.me comes from lack of schema information. For example, in Zhishi.me there is no definition whether a property is an object property or a datatype property. Thus, misuse of these two kinds of properties cannot be detected simply by SPARQL queries, which can be used in DBpedia to detect the same problems.

# 3    Analysis on DBpedia and Zhishi.me

In this section, we present problems based on the analysis of DBpedia and Zhishi.me. For each problem, we will give suitable examples to illustrate them, and discuss how they could be identified.

## 3.1    Using Different IRIs to Represent the Same Property

Since resources from Chinese enclyclopedias are created manually by various editors, there may exist some nonstandard editions of resources, such as unnecessary characters in the identifier of an entity. For example, in baidubaike,

---

[1] http://athena.ics.forth.gr:9090/RDF/VRP/

"baidu:shengao" and "baidu:?shengao" have the same semantics, because both of them are used to describe the height of a person. However, they are treated by machine as different properties. Such cases will increase the redundancy of knowledge base and reasoning complexity.

Our procedure for the detection of properties which have the same meaning is given as follows:

**Step 1: Property acquisition.** First we collect the names of all the properties by SPARQL queries. We assume that every predicate in an RDF triple is a property.

**Step 2: Normalization.** We extract English letters, Chinese characters, and numeric characters from every property name. Besides, all the English letters are transformed into their lowercase. For example, "?shengao" is normalized to "shengao", and "ISBN" is normalized to "isbn".

**Step 3: String segmentation.** The main purpose of this step is to divide a string into meaningful units. We used IKAnalyser [2], an open source Chinese text segmentation toolkit based on java, as the segmentation tool. For every normalized property name, the segmentation result is a series of meaningful units seperated by blank spaces. In order to improve efficiency, units from every segmentation are reordered lexically. In that case, the segmentation result of "xuexingshengao" is the same as that of "shengaoxuexing".

**Step 4: String similarity computation.** We take Levenshtein Distance as the metric for measuring the similarity between two properties that have been processed by the above steps. The greater the Levenshtein Distance, the smaller the similarity. For strings $s$ and $t$, their similarity is given by

$$sim(s,t) = 1 - \frac{ld(s,t)}{(s.length, t.length)}$$

where $ld(s,t)$ refers to the Levenshtein Distance of $s$ and $t$.

Two properties $s$ and $t$ having the same semantics would lead to $sim(s,t) = 1.0$.

## 3.2 Inverse Definition of Subclass Relations

A subclass-of relation states that all the instances of one class are instances of another [3]. That is, axiom A $\sqsubseteq$ B states that the set of individuals in class A is a subset of the set of individuals in class B. Through our analysis, we found some subclass-of relations which were defined inversely.

It is reasonable to assume that if class A is a subclass of class B, then B has more individuals than A, and they should have some common individuals. Thus, if axiom A $\sqsubseteq$ B is defined inversely, the following two conditions will be satisfied:

*Condition*1: $n_1 > n_2$, where $n_1$ and $n_2$ refer to the number of individuals belonging to A and B respectively.

---

[2] https://code.google.com/p/ik-analyzer/
[3] http://www.w3.org/TR/rdf-schema/

*Condition*2: $n_{11} > 0$, where $n_{11}$ represents the number of individuals belonging to both A and B.

We should notice that the aforementioned conditions do not always work, because when editing a resource in an encyclopedia, the editor may not give all the classes it belongs to. In order to improve accuracy, we will use Wikipedia to check the correctness of $A \sqsubseteq B$.

From the web page of category A, we can get the set of its parent categories $S_A$. Figure 1 give an example to show how to get the parent categories for the Chinese version of category "Education_in_Africa".

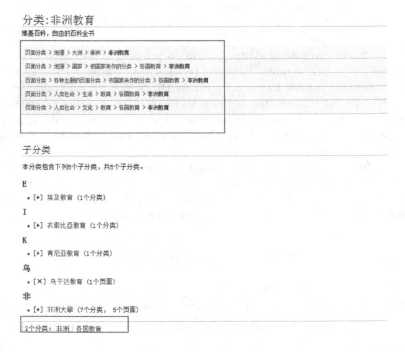

**Fig. 1.** Web page of the Chinese version of category "Education_in_Africa"

All the categories which appear in the red squares in Figure 1 are the parent categories of the present category, so we add each of them to the set of its parent categories without repetition. Classes in the upper squares illustrate pathes from the top class to the present class, while classes at the bottom of the page illustrate categories the present class can be divided into. In this example we can have

$$S_{非洲教育} = \{页面分类, 地理, 大洲, 非洲, 国家, 依国家来作的的分类, 各国教育,$$
$$各种主题的页面分类, 人类社会, 生活, 教育, 文化\}.$$

We propose Algorithm 1 to check the correctness of $A \sqsubseteq B$. Algorithm 1 takes the relation to be checked as input. The output of the algorithm may have three possible values: right, inverse, undetermined. "right" means the input relation is right, "inverse" means the input relation has been defined inversely, and "undetermined" means the correctness of the relation cannot be determined. We use the following schema to explain Algorithm 1.

**IF** both A and B are categories in Wikipedia, **THEN** (line 1)

**IF** $S_A$ contains B, **THEN** A $\sqsubseteq$ B is right (i.e., value "right" is returned). (line 2-line 4)

**ELSE IF** A is an elements of $S_B$, **THEN** A $\sqsubseteq$ B is defined inversely (i.e., value "inverse" is returned). (line 6-line 8)

**ELSE IF** neither A is an element of $S_B$ nor B is an elements of $S_A$, **THEN** the correctness of A $\sqsubseteq$ B cannot be determined in Wikipedia (i.e., value "undetermined" is returned). (line 10)

**IF** A or B is not a category in Wikipedia, **THEN** (line 14)

**IF** B is a category and A is a resource belongs to B, **THEN** A $\sqsubseteq$ B is right. (line 15-line 16)

**ELSE IF** A is a category and B is a resource belongs to A, **THEN** A $\sqsubseteq$ B is inverse.(line 18-line 19)

**ELSE** the correctness of A $\sqsubseteq$ B cannot be determined. (line 21)

Here we give an example to illustrate Algorithm 1.

For axiom `Environmental_protection` $\sqsubseteq$ `Low-carbon`, we first search for the category "Environmental_protection" in Wikipedia and get the set of all its parent categories.

$$S_{Environmental\_protection}=\{\text{Contents, Society, Environment, Human\_rights\_by\_issue}\}$$

"Low-carbon" is not an element of $S_{Environmental\_protection}$. Then we search for category "Low-carbon" in Wikipedia and found there is no such a category. So we check resource "Low-carbon" in Wikipedia. From the returned web page we can see that this resource belongs to category "Environmental_protection". So axiom `Environmental_protection` $\sqsubseteq$ `Low-carbon` is defined inversely.

To summarize, the whole detection process can be divided into two steps. First we get all subclass relations in which the parent class has less individuals than its subclass, and the two classes have common individuals, then we check their correctness in Wikipedia.

### 3.3  Class and Individual Belonging to It Share the Same Identifier

The distinction between a class and an individual has been discussed in W3C Recommendation for OWL [4]. A class is simply a name of a set, and the elements in the set are its instances. That is to say, individuals refer to actual entities in the world and classes correspond to sets of those real world entities. In some languages like OWL DL, classes are not allowed to be used as instances.

However, in Zhishi.me, some entities are used both as classes and as instances. There are two reasons for such phenomenon. The first one is that in some cases a class can be considered an instance of another class. The other one comes from

---

[4] `http://www.w3.org/TR/owl-guide/`

**Algorithm 1.** Check the correctness of A ⊑ B

**Input:**
A ⊑ B: the subclass-of relation to check
**Output:**
truth: the correctness of the input relation

1: **if** both A and B are categories in Wikipedia **then**
2:     $S_A$={all the parent categories of A}
3:     **if** B is an element of $S_A$ **then**
4:         truth=right;
5:     **else**
6:         $S_B$={all the parent categories of B}
7:         **if** A is an element of $S_B$ **then**
8:             truth=inverse;
9:         **else**
10:            truth=undetermined;
11:        **end if**
12:    **end if**
13: **end if**
14: **if** A or B is not a category of Wikipedia **then**
15:    **if**  B is a category and A is a resource belongs to B **then**
16:        truth=right
17:    **else**
18:        **if** A is a category and B is a resource belongs to A **then**
19:            truth=inverse;
20:        **else**
21:            truth=undetermined;
22:        **end if**
23:    **end if**
24: **end if**
    **return** truth;

the neglect of the differences between instance-of relationship and subclass-of relationship.

Through our analysis on DBpedia and Zhishi.me, we found blurred usage of classes and individuals, which cannot be explained by the above two ways. That is, an entity is used both as a class and as an instance belonging to it.

This problem can be identified by a SPARQL query:

```
PREFIX rdfs: <http://www.w3.org/2000/01/rdf-schema#>
SELECT DISTINCT ?x ?y WHERE
{?x rdfs:label ?label1.
?y rdfs:label ?label2.
?x a ?y.
FILTER(?label1=?label2).}
```

In DBpedia, we found only one such case, that is

<http://dbpedia.or/resource/Animal> a <http://dbpedia.org/ontology/Animal>

In the above triple, the subject has the same identifier as the object. It asserts that Animal is an individual of class Animal, which is rather obscure and hard to explain.

However, in hudongbaike, we found 6,731 such classes, which constituted about 27% of all the classes in hudongbaike.

We did analysis on the data returned by the above query and found that there were three ways in which class and individual were used improperly: (1) An entity which should be a class was also defined as an individual. (2) An entity which is an individual was also defined as a class. (3) The identifier of an entity was ambiguous. For example, in hudongbaike, "ribendaxue" is defined as an individual of class "ribendaxue". When the entity "ribendaxue" is used as an individual, it refers to Nihon University, which is an university in Japan. However, when it is used as a class, it refers to the set of all the universities in Japan.

## 3.4 Misuse of Properties

### 3.4.1 Misuse of skos:broader and skos:narrower

In Zhishi.me, skos:broader and skos:narrower are used to represent class hierarchy. In convention, these two properties are used to represent a direct hierarchical link between two concepts, so they are not declared to be transitive. We can use SPARQL queries to determine whether the two relations are used correctly. Let us take skos:broader as an example, the misuse of skos:broader can be detected by a non-empty result set for the SPARQL query

```
PREFIX skos: <http://www.w3.org/2004/02/skos/core#>
SELECT DISTINCT(*) WHERE
{?x skos:broader ?y.
?y skos:broader ?z.
?x skos:broader ?z.}
```

By applying the above query in baidubake and hudongbaike, we get non-empty result sets. That is to say, skos:braoder is misused in the two encyclopedias. For example, among the results returned from baidubaike, we can get the following triples from which we can see skos:broader is used incorrectly:

*baidu:Singapore skos:broader baidu:Countries_in_Asia*
*baidu:Countries_in_Asia skos:broader baidu:Countries*
*baidu:Singapore skos:broader baidu:Countries*

In DBpedia, class hierarchy is represented by rdfs:subClassOf, which is a transitive property. Thus this problem would not happen in DBpedia.

### 3.4.2    Misuse of Datatype Properties and Object Properties

Datatype properties are used to represent relations between instance and RDF literal, while object properties represent relations between two instances. These two types of properties can be distinguished by terms owl:DatatypeProperty and owl:ObjectProperty. However, in reality, their difference may be neglected.

In DBpedia, since every property has been defined as either an object property or a datatype property, misuse of properties can be detected by SPARQL queries. For example, misuse of datatype properties can be detected by a nonempty set for the SPARQL query:

```
PREFIX owl: <http://www.w3.org/2002/07/owl#>
SELECT DISTINCT ?p WHERE
{?s ?p ?o.
?p a owl:DatatypeProperty.
FILTER isIRI(?o).}
```

For Zhishi.me, since there is no definition whether a property is a datatype property or an object property, the above queries for detection is not applicable. However, we did find some properties that were used both as datatype properties and as object properties. For example,

> *baidu:*代码重构（*VisualBasic*版）  *baidu:*原版名称 *"Professional Refactoring in Visual Basic"@zh*
> *baidu:*通信经济学  *baidu:*原版名称  *baidu:*通信经济学

The predicate in the above triples is used both as a datatype property and as an object property.

Here we give a method for the detection of properties which are used both as datatype property and object property in Zhishi.me.

**Step 1:** Use the following query to get all the properties used as object property.

```
SELECT DISTINCT ?p WHERE
{?s ?p ?o.
FILTER isIRI(?o).}
```

**Step 2:** For every property returned in Step 1, check whether it takes literals as objects or not.

By following thess steps in baidubaike, 3,817 object properties were selected out in Step 1(properties coming from other vocabularies such as skos:broader, rdf:type were excluded). After Step 2, we finally got 274 properties which were used both as object property and datatype property.

# 4    Experiments

For errors that can be detected by SPARQL queries, we have demonstrated the validity of the SPARQL query in the previous subsections. Hence, in this section we only report the results of experiments which are conducted for validating the proposed semi-automatic methods.

## 4.1    Detection for IRIs Representing the Same Property

We report the experimental results conducted on baidubaike.

We got 10,653 properties from baidubaike. We assume that properties $s$ and $t$ having the same semantics would lead to $sim(s, t) = 1.0$. Following the afore-mentioned procedure in Section 3.1, we got 117 such property pairs.

Since there is no standard for our experiments, we checked every property pair manually to see if they really represent the same semantics. Through the manual checking, we only found three property pairs which were classified incorrectly. The reason is that these six properties all contain more than one meaningful units, and different orders of the units will lead to different meanings. For example, "zazhizhubian" refers to the chief editor of a magazine, while "zhubianzazhi" refers to main magazines a person works on. For these pair of properties, our method incorrectly classifies them into the same meaning.

We did analysis on the experimental results and found that this type of problem mainly came from three aspects:

(1) Some properties contain meaningless characters. These characters may come from the unexpected input during the edition of a web page. For example, "2004– –" is a property which contains meaningless characters "– –".

(2) As to English abbreviations, some properties use their lowercase, while some properties capitalize all the English letters. For example, we found "GDP" and "gdp" have the same meaning.

(3) Some properties contain more than one meaningful units. These units are ordered differently in different properties. In that case we need to segment them and reorder the units. "xuexingxingzuo" and "xingzuo/xuexing" is one such example.

## 4.2    Detection for Inversely Defined Subclass-of Relations

We did experiments to see whether we can find inversely defined subclass-of relations by using the methods we have proposed in Section 3.2. We only considered the subclass-of relations where the number of common individuals between two classes is no less than 10. Our experiments were conducted on DBpedia and hudongbaike.

From DBpedia we got 316 subclass-of relations. After counting the individual number of every class in those relations, we found only two relations which satisfied $Condition1$ and $Condition2$: Settlement $\sqsubseteq$ PopulatedPlace and Eukaryote $\sqsubseteq$ Animal . Next, we checked them in Wikipedia and found that the correctness of `Settlement` $\sqsubseteq$ `PopulatedPlace` could not be determined, while `Eukaryote` $\sqsubseteq$ `Animal` was inversely defined.

In hudongbaike, we got 70,107 subclass-of relations which were defined by skos:broader (as discussed before, skos:broader was incorrectly used. Here we just regarded skos:broader as rdfs:subClassOf). With our detection methods, we got 644 relations that satisfy *Condition*1 and *Condition*2. We then checked every relation in the Chinese version of Wikipedia. Among the 644 relations, we found 12 of them that were inversely defined. Besides, there were about half of those 644 relations whose correctness could not be determined. This is because classes involved in those relations were not categories in Wikipedia.

Note that, our detection in wikipedia is based on strict string match. This will lead to a relatively smaller number of relations that can be determined as correct. For example, in hudongbaike, class "ChongQing" is defined as the subclass of "Municipalities_of_China". However, in wikipedia, "ChongQing" is a subclass of "Municipalities_of_People's_Republic_of_China", and class "Municipalities_of_China" does not exist in Wikipedia. So the subclass-of relation `ChongQing` $\sqsubseteq$ `Municipalities_of_China` is undetermined. It is obvious that "Municipalities_of_People's_Republic_of_China" and "Municipalities_of_ China" refer to the same class. However, due to the use of strict string match, the correctness of relation `ChongQing` $\sqsubseteq$ `Municipalities_of_China` is undetermined.

## 5   Conclusion

Error detection plays a very important role in improving the quality of Linked Data. In this paper we identified some problems which may occure in DBpedia and Zhishi.me. For each type of error, we either give SPARQL queries for automatic detection or present semi-automatic detection methods. We did experiments on DBpedia and Zhishi.me. Our experimental results show that the proposed methods can detect analyzed problems.

Future work includes a deeper analysis on RDF datasets to find out more problems. We may conduct experiments on other cross-domain RDF datasets such as Freebase, Yago, and so on. Moreover, we will proposed methods to deal with those errors.

## References

1. Auer, S., Bizer, C., Kobilarov, G., Lehmann, J., Cyganiak, R., Ives, Z.: DBpedia: A nucleus for a web of open data. In: Aberer, K., et al. (eds.) ISWC/ASWC 2007. LNCS, vol. 4825, pp. 722–735. Springer, Heidelberg (2007)
2. Bizer, C., Heath, T., Berners-Lee, T.: Linked data-the story so far. Int. J. Semantic Web Inf. Syst. 5(3), 1–22 (2009)
3. Fleischhacker, D., Völker, J.: Inductive learning of disjointness axioms. In: Meersman, R., et al. (eds.) OTM 2011, Part II. LNCS, vol. 7045, pp. 680–697. Springer, Heidelberg (2011)
4. Hogan, A., Harth, A., Passant, A., Decker, S., Polleres, A.: Weaving the pedantic web. In: LDOW (2010)

5. Hogan, A., Umbrich, J., Harth, A., Cyganiak, R., Polleres, A., Decker, S.: An empirical survey of linked data conformance. J. Web Sem. 14, 14–44 (2012)
6. Lehmann, J., Bühmann, L.: ORE-a tool for repairing and enriching knowledge bases. In: Patel-Schneider, P.F., Pan, Y., Hitzler, P., Mika, P., Zhang, L., Pan, J.Z., Horrocks, I., Glimm, B. (eds.) ISWC 2010, Part II. LNCS, vol. 6497, pp. 177–193. Springer, Heidelberg (2010)
7. Niu, X., Sun, X., Wang, H., Rong, S., Qi, G., Yu, Y.: Zhishi.me - weaving chinese linking open data. In: Aroyo, L., Welty, C., Alani, H., Taylor, J., Bernstein, A., Kagal, L., Noy, N., Blomqvist, E. (eds.) ISWC 2011, Part II. LNCS, vol. 7032, pp. 205–220. Springer, Heidelberg (2011)
8. Péron, Y., Raimbault, F., Ménier, G., Marteau, P.F., et al.: On the detection of inconsistencies in rdf data sets and their correction at ontological level. In: ISWC (2011)
9. Rahm, E., Do, H.H.: Data cleaning: problems and current approaches. J. IEEE Data Eng. Bull. 23(4), 3–13 (2000)
10. Töpper, G., Knuth, M., Sack, H.: DBpedia ontology enrichment for inconsistency detection. In: I-SEMANTICS, pp. 33–40 (2012)
11. Völker, J., Niepert, M.: Statistical schema induction. In: Antoniou, G., Grobelnik, M., Simperl, E., Parsia, B., Plexousakis, D., De Leenheer, P., Pan, J. (eds.) ESWC 2011, Part I. LNCS, vol. 6643, pp. 124–138. Springer, Heidelberg (2011)
12. Wang, R.Y., Strong, D.M.: Beyond accuracy: what data quality means to data consumers. J. Management Inform. Systems 12(4), 5–33 (1996)

# Multi-criteria Axiom Ranking Based on Analytic Hierarchy Process

Jianfeng Du, Rongfeng Jiang, and Yong Hu

Guangdong University of Foreign Studies, Guangzhou 510006, China
jfdu@gdufs.edu.cn

**Abstract.** Axiom ranking plays an important role in ontology repairing. There has been a number of criteria that can be used in axiom ranking, but there still lacks a framework for combining multiple criteria to rank axioms. To provide such a framework, this paper proposes an analytic hierarchy process (AHP) based approach. It expresses existing criteria in a hierarchy and derives weights of criteria from pairwise comparison matrices. All axioms are then ranked by a weighted sum model on all criteria. Since the AHP based approach does not work when a pairwise comparison matrix is insufficiently consistent, a method is proposed to adjust the matrix. The method expresses the adjustment problem as an optimization problem solvable by level-wise search. To make the proposed method more practical, an approximation of it is also proposed. Experimental results show that the proposed method is feasible for small pairwise comparison matrices but is hard to scale to large ones, while the approximate method scales well to large pairwise comparison matrices.

## 1 Introduction

Since the W3C recommended the standard Web Ontology Language (OWL) for modeling ontologies, more and more ontologies have been developed in different domains. During the life cycle of an ontology, it is hard to guarantee that the ontology is perfect and has not any problem, thus ontology repairing [12] is often needed. It revises or removes axioms from a problematic ontology to make it suitable for subsequent processing.

There are two kinds of ontology repairing, namely explicit repairing and implicit repairing. Explicit repairing needs to materialize the repaired result before subsequent processing. It has been applied to correct unsatisfiable concepts [12], revise problematic axioms [13,7], and select a consistent subset of the ontology to perform subsequent reasoning [10,18]. Axiom ranking is employed in these applications to determine an order for revising or removing axioms. Implicit repairing does not really materialize the repaired result but directly infers what follows from the repaired result. It is often applied in some inconsistency-tolerant reasoning mechanisms, such as the lexicographic inference mechanism studied in [5] and the weight-based consistent query answering mechanism proposed in [6], for which the semantics are defined over possible repaired results but repairing

G. Qi et al. (Eds.): CSWS 2013, CCIS 406, pp. 118–131, 2013.

are not really done. For these mechanisms, axiom ranking is also required to determine the priorities or weights of axioms.

Although axiom ranking is crucial in both kinds of ontology repairing, the research work on axiom ranking is insufficient. Most existing work focuses on a single criterion for axiom ranking. The single criterion can be either a syntactic one, such as the usage or provenance information [12], or a semantic one, such as the frequency in minimal unsatisfiable subsets [12], the impact on entailments or test cases [12,13], the degree of inconsistency [4,11,16], and the degree of incoherency [17]. Different criteria for ranking axioms may not be coherent and probably have different suitable scenarios. Combining multiple criteria is a natural way to avoid the bias caused by a single criterion. However, there is little work on combining multiple criteria to rank axioms. As far as we know, there is only one attempt which uses the weighted sum of scores of different criteria where weights are assigned by users [12]. Although this way is easily implemented, there is no method provided to acquire reasonable weights.

In this paper, we propose an analytic hierarchy process (AHP) [19] based approach to ranking axioms. AHP is a multi-criteria decision making framework widely adopted in various domains [20]. This AHP based approach exploits a weighted sum model for ranking axioms, where weights are automatically derived from pairwise comparison matrices based on the importance of each criterion. Since the setting of relative importance of a criterion over another one is easier and more accurate than the direct setting of weights, AHP gives a reasonable way to acquire weights of different criteria. Moreover, AHP expresses criteria in a hierarchy. Only all sibling criteria (i.e. criteria with the same parents) need to be compared. This way alleviates the effort to pairwise compare criteria.

In AHP the users need to manually construct a pairwise comparison matrix $A = (a_{ij})_{n \times n}$ for each group of sibling criteria $C_1, ..., C_n$. A judgement $a_{ij}$ in the matrix represents the relative importance of $C_i$ over $C_j$. A perfect matrix should be *reciprocal*, namely $a_{ij} = 1/a_{ji}$ for all $i$ and $j$, and *consistent*, namely $a_{ij}a_{jk} = a_{ik}$ for all $i, j$ and $k$. While it is easy for users to guarantee a matrix to be reciprocal, it is hard to guarantee consistency. AHP allows a certain level of inconsistency, which is measured by the *consistency ratio* (*CR*) [19]. CR is a global characteristic of the pairwise comparison matrix. When CR is less than 0.1, the inconsistency is deemed to be relatively small and the matrix is said to be *sufficiently consistent*. Only when a pairwise comparison matrix is sufficiently consistent can it be used to derive reasonable weights of criteria.

It is still not easy to construct a sufficiently consistent pairwise comparison matrix, hence we propose an automatic method for adjusting the judgements in the matrix to render it sufficiently consistent. The method expresses the adjustment problem as an optimization problem solvable by level-wise search. To make the method more practical, we also apply the local search paradigm to approximate the method. Experimental results on randomly generated pairwise comparison matrices show that the proposed method is feasible for small pairwise comparison matrices but is hard to scale to large ones, while the approximate method scales well to large pairwise comparison matrices.

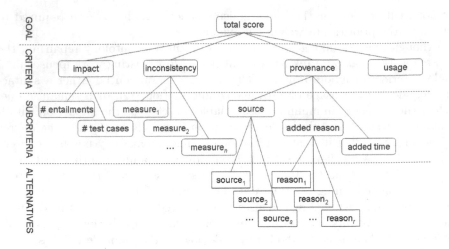

**Fig. 1.** A general criterion hierarchy for ranking axioms

There are two main contributions in this paper. On the one hand, an AHP based approach is proposed to combine multiple criteria for ranking axioms. Existing criteria are summarized in a hierarchy, providing a framework for ranking axioms. On the other hand, two algorithms for adjusting judgements in an insufficiently consistent pairwise comparison matrix, where one is exact and the other is approximate and scalable, are proposed to enable reasonable weights of criteria to be derivable from the matrix.

## 2    An AHP Based Approach to Axiom Ranking

Axiom ranking can be treated as a multi-criteria decision making (MCDM) problem. The AHP framework deals with MCDM problems by determining the weights of criteria through pairwise comparison of the criteria. The pairwise comparison matrix expresses the relative importance of each criterion over others. The procedure for ranking axioms is summarized as follows:

1. Construction of the criterion hierarchy.
2. Construction of pairwise comparison matrices.
3. Derivation of weights of criteria from a pairwise comparison matrix.
4. Computation of the total score for each axiom.

### 2.1    Construction of the Criterion Hierarchy

A general criterion hierarchy for ranking axioms has been developed and is shown in Fig. 1. These criteria are selected by investigating the common features of existing research work [12,13,4,17,11,16]. Totally four general criteria are identified and defined: impact, inconsistency, provenance and usage. In addition, sub-criteria and alternatives are added as descendants of these criteria.

| | source$_1$ | source$_2$ | source$_3$ |
|---|---|---|---|
| source$_1$ | 1 | 3 | 5 |
| source$_2$ | 1/3 | 1 | 3 |
| source$_3$ | 1/5 | 1/3 | 1 |

| | reason$_1$ | reason$_2$ | reason$_3$ |
|---|---|---|---|
| reason$_1$ | 1 | 5 | 3 |
| reason$_2$ | 1/5 | 1 | 1/4 |
| reason$_3$ | 1/3 | 4 | 1 |

**Fig. 2.** Two example PCMs for different sources and different added reasons

The *impact* criterion represents the impact on entailments or test cases [12,13]. Two sub-criteria whose values are numeric are added as children, namely the number of impacted entailments and the number of impacted test cases.

The *inconsistency* criterion represents the degree of inconsistency [4,11,16] or the degree of incoherence [17]. Different sub-criteria whose values are numeric are added as children, each of which is a certain measure of the degree of inconsistency/incoherence. The frequency of an axiom in minimal unsatisfiable subsets [12] is also treated as such a measure.

The *provenance* criterion represents the provenance information [12]. Different sub-criteria that correspond to different aspects of the provenance information are added as children: the *source* sub-criterion represents the reliability of the source; the *added reason* sub-criterion represents the reason or context for which the axiom was added; the *added time* sub-criterion represents the time the axiom was added. Since the score of the source (resp. added reason) sub-criterion depends on categorical values, different alternatives that correspond to different sources (resp. different reasons or contexts) are added as children of the source (resp. added reason) sub-criterion. The alternative level differs from other levels in that only one out of all sibling alternatives is applicable to a certain axiom, whereas all sibling (sub-)criteria are applicable to a certain axiom.

The *usage* criterion determines how often entities in the signature of the axiom have been referenced in other axioms [12]. The score of this criterion is numeric and can be directly computed, thus no children of it is added.

The above hierarchy can be adapted or simplified for concrete applications.

### 2.2 Construction of Pairwise Comparison Matrices

A pairwise comparison matrix (PCM) $A = (a_{ij})_{n \times n}$ is constructed for each group of sibling criteria $C_1, ..., C_n$, where sub-criteria and alternatives are also treated as criteria, and $a_{ij}$ is the relative importance of criterion $C_i$ over criterion $C_j$, which should be in the Saaty scale [19], i.e., $a_{ij} \in \{1/9, ..., 1/2, 1, 2, ..., 9\}$. It is also required that a PCM be reciprocal, i.e., $a_{ij} = 1/a_{ji}$ for all $i$ and $j$. All PCMs are specified by users. Two example PCMs for children of score and children of added reason in the criterion hierarchy for ranking axioms (see Fig. 1), respectively, are given in Fig. 2.

### 2.3 Derivation of Weights of Criteria from a Comparison Matrix

The weight $w_i$ of each criterion $C_i$, derived from the PCM $A = (a_{ij})_{n \times n}$ on criteria $C_1, ..., C_n$, should be such that $\sum_{i=1}^{n} w_i$ equals to 1 and $w_i/w_j$ is close to $a_{ij}$. There are two widely used methods to derive the weight of each criterion from a PCM. One is eigenvector and the other is geometric mean. These two methods derive the same weights when the PCM is consistent, i.e., $a_{ij}a_{jk} = a_{ik}$ for all $i, j$ and $k$. In this paper, the geometric mean method is used because it overcomes the problem of left-right eigenvector asymmetry in the eigenvector method [2]. This method is simple and $w_i$ is defined as the normalized geometric mean of $\{a_{ij}\}_{1 \le j \le n}$:

$$w_i' = (\prod_{j=1}^{n} a_{ij})^{1/n} \quad \text{for } i = 1, ..., n \tag{1}$$

$$w_i = w_i' / \sum_{j=1}^{n} w_j' \quad \text{for } i = 1, ..., n \tag{2}$$

Due to the Saaty scale used in a PCM, a PCM is usually inconsistent. Moreover, the manual construction of a PCM cannot guarantee consistency either. Fortunately, a certain level of inconsistency is acceptable in practice. The level of inconsistency is measured by the *consistency ratio* (*CR*) [19]. CR is defined as the ratio of the *consistency index* (*CI*) to the *random consistency index* (*RI*):

$$\lambda_{max} = \frac{1}{n} \sum_{i=1}^{n} \frac{\sum_{j=1}^{n} a_{ij} w_j}{w_i} \tag{3}$$

$$CI = \frac{\lambda_{max} - n}{n - 1} \tag{4}$$

$$CR = \frac{CI}{RI(n)} \tag{5}$$

In (3), $\lambda_{max}$ is an approximation of the principal eigenvalue of $A = (a_{ij})_{n \times n}$. In (5), $RI(n)$ is the random consistency index for matrices of order $n$, which is the average value of CIs for random reciprocal matrices of order $n$, obtained by a large number of random tests. A list of $RI(n)$ for different $n \ge 3$ [1] is shown below. A reciprocal matrix of order $n \le 2$ must be consistent, thus there is no value of $RI(n)$ for $n \le 2$.

| $n$ | 3 | 4 | 5 | 6 | 7 | 8 | 9 | 10 | 11 | 12 | 13 | 14 | 15 |
|---|---|---|---|---|---|---|---|---|---|---|---|---|---|
| $RI(n)$ | 0.525 | 0.882 | 1.115 | 1.252 | 1.341 | 1.404 | 1.452 | 1.484 | 1.513 | 1.535 | 1.555 | 1.570 | 1.583 |

When CR is less than 0.1, the inconsistency of a PCM is deemed to be acceptable [19] and we call this PCM *sufficiently consistent*. In practice, the weights derived from a PCM can be used only when the PCM is sufficiently consistent. Take the two PCMs given in Fig. 2 for example, the CR of the left PCM is 0.037 while the CR of the right PCM is 0.082, thus both PCMs are sufficiently consistent.

## 2.4    Computation of the Total Score for Each Axiom

The total score of an axiom can be computed recursively from leafs to the root in the criterion hierarchy (Fig. 1). Given an axiom, the score attached to every leaf is computed first. In particular, the score attached to an alternative is set as 1 if the axiom has this alternative, or set as 0 if the axiom has a different one. According to the relativity nature of weights, the score attached to every node in the hierarchy should have a uniform scale. To this end we normalize the score attached to a leaf to a real number between 0 and 1. For a node whose children are alternatives $C_1, ..., C_n$, the score attached to it is defined as $w_k / \max\{w_i\}_{1 \leq i \leq n}$, where $w_i$ is the weight of $C_i$ $(1 \leq i \leq n)$ and $C_k$ is the alternative that the given axiom has. The division by $\max\{w_i\}_{1 \leq i \leq n}$ ensures that the score is up to 1. For a node whose children are criteria or sub-criteria $C_1, ..., C_n$, the score attached to it is defined as $\sum_{i=1}^{n} s_i w_i$, where $w_i$ is the weight of $C_i$ and $s_i$ is the score attached to $C_i$ $(1 \leq i \leq n)$. This score is also in the range between 0 and 1 because $\sum_{i=1}^{n} w_i = 1$ and $0 \leq s_j \leq 1$ for all $1 \leq j \leq n$.

*Example 1.* Suppose in Fig. 1 the added reason sub-criterion has two children, the source sub-criterion has three children, and the inconsistency criterion has four children. Moreover, suppose the weights of all criteria, derived from certain PCMs, are given below.

| reason$_1$ | reason$_2$ | source$_1$ | source$_2$ | source$_3$ | measure$_1$ | measure$_2$ | measure$_3$ | measure$_4$ |
|---|---|---|---|---|---|---|---|---|
| $w_1$ | $w_2$ | $w_3$ | $w_4$ | $w_5$ | $w_6$ | $w_7$ | $w_8$ | $w_9$ |
| # entail | # test | source | added reason | added time | impact | inconsistency | provenance | usage |
| $w_{10}$ | $w_{11}$ | $w_{12}$ | $w_{13}$ | $w_{14}$ | $w_{15}$ | $w_{16}$ | $w_{17}$ | $w_{18}$ |

Given an axiom, suppose all scores attached to leafs in the criterion hierarchy have been normalized and are given below.

| reason$_1$ | reason$_2$ | source$_1$ | source$_2$ | source$_3$ | measure$_1$ | measure$_2$ | measure$_3$ | measure$_4$ |
|---|---|---|---|---|---|---|---|---|
| 1 | 0 | 0 | 1 | 0 | $s_6$ | $s_7$ | $s_8$ | $s_9$ |
| # entail | # test | source | added reason | added time | impact | inconsistency | provenance | usage |
| $s_{10}$ | $s_{11}$ | — | — | $s_{14}$ | — | — | — | $s_{18}$ |

Then, the score attached to source is $w_4 / \max\{w_3, w_4, w_5\}$, the score attached to added reason is $w_1 / \max\{w_1, w_2\}$, the score attached to impact is $s_{10} w_{10} + s_{11} w_{11}$, the score attached to inconsistency is $s_6 w_6 + s_7 w_7 + s_8 w_8 + s_9 w_9$, the score attached to provenance is $w_4 w_{12} / \max\{w_3, w_4, w_5\} + w_1 w_{13} / \max\{w_1, w_2\} + s_{14} w_{14}$, and the total score of the given axioms is

$$(s_{10} w_{10} + s_{11} w_{11}) w_{15} + (s_6 w_6 + s_7 w_7 + s_8 w_8 + s_9 w_9) w_{16} +$$
$$\left( \frac{w_4 w_{12}}{\max\{w_3, w_4, w_5\}} + \frac{w_1 w_{13}}{\max\{w_1, w_2\}} + s_{14} w_{14} \right) w_{17} + s_{18} w_{18} .$$

## 3   A Method for Adjusting Pairwise Comparisons

A pairwise comparison matrix (PCM) constructed by users may not be sufficiently consistent. It is needed to adjust an insufficiently consistent PCM before using it to derive weights of criteria. However, the manual adjustment is time-consuming and error-prone, hence (semi-)automatic methods are required. Some methods adopt an interactive mode, each time identifying and modifying the most inconsistent judgement in the PCM while trying to preserve the remaining comparison judgements [14,3,8]. These methods still need some human efforts and may not quickly resolve the inconsistency. Other methods reduce the inconsistency by an auto-adaptive process, which adjusts the entire PCM but does not keep comparison judgements still in the Saaty scale [21]. All the above methods do not guarantee that the adjusted PCM is as close as possible to the original PCM.

We consider automatically adjusting the judgements in a PCM. The adjusted PCM should have the following two properties. First, it is as close as possible to the original PCM. This is desirable because a PCM used in AHP is carefully specified by users and it should not be modified too much. Second, all adjusted judgements are in the Saaty scale. This can guarantee the intelligibility of the adjusted PCM since each number in the Saaty scale has a standard interpretation but numbers outside have not. To make an adjusted PCM have the above properties, we propose a method which expresses the adjustment problem as an optimization problem. We assume that the original PCM is reciprocal, which can easily be guaranteed by users. Since any two adjacent numbers in the Saaty scale has the same cognitive distance, we map a number $t$ in the Saaty scale to an integer $f(t)$ and define the distance between two numbers $t_1$ and $t_2$ in the Satty scale as $|f(t_1) - f(t_2)|$, where $f(t)$ is defined below.

$$f(t) = \begin{cases} t - 1 & t \in \{1, 2, ..., 9\} \\ 1 - \frac{1}{t} & t \in \{\frac{1}{9}, \frac{1}{8}, ..., \frac{1}{2}\} \end{cases} \qquad (6)$$

We then define the distance between two PCMs of the same order as the sum of distances between corresponding judgements in the two PCMs. Let $A = (a_{ij})_{n \times n}$ be the original PCM, and $X = (x_{ij})_{n \times n}$ be an adjusted PCM which is reciprocal and sufficiently consistent. Under the assumption that $A$ is reciprocal, the distance between $A$ and $X$ is double of the sum of distances between corresponding judgements in the upper triangular parts of the two PCMs. Hence, the object function in the adjustment problem is the following minimization function.

$$\min \sum_{i=1}^{n-1} \sum_{j=i+1}^{n} |f(x_{ij}) - f(a_{ij})| \qquad (7)$$

Let $CR((a_{ij})_{n \times n})$ be a function returning the consistency ratio of $(a_{ij})_{n \times n}$, defined by (1)–(5). Since $X$ should be sufficiently consistent, the following constraint is added to the adjustment problem.

$$CR((x_{ij})_{n \times n}) < 0.1 \qquad (8)$$

Since $X$ should be reciprocal, the following constraints are also added to the adjustment problem.

$$x_{ii} = 1 \qquad \text{for } i = 1, ..., n \qquad (9)$$

$$x_{ji} = 1/x_{ij} \qquad \text{for } i = 1, ..., n-1 \text{ and for } j = i+1, ..., n \qquad (10)$$

Finally, users may specify the range of some judgements in $X$. Let $S_{ij}$ be the set of numbers in the Saaty scale that $x_{ij}$ should be in. $S_{ij}$ is defined as the default set $\{1/9, ..., 1/2, 1, 2, ..., 9\}$ if it is not specified by users. We assume that

**Algorithm 1.** SearchOptimalPCM$((a_{ij})_{n \times n}, \{S_{ij} \mid 1 \leq i \leq n-1, i+1 \leq j \leq n\})$

**Input:** A reciprocal and insufficiently consistent PCM $(a_{ij})_{n \times n}$ and a collection of sets of allowed numbers $\{S_{ij} \mid 1 \leq i \leq n-1, i+1 \leq j \leq n\}$ for judgements in an adjusted PCM.

**Output:** An adjusted PCM or the message "no solution".

1:  $d_{\max} \leftarrow \sum_{i=1}^{n-1} \sum_{j=i+1}^{n} \max\{|f(t) - f(a_{ij})| \mid t \in S_{ij}\};$
2:  **for** $d$ from 1 to $d_{max}$ **do**
3:     **for** each candidate $(c_{ij})_{1 \leq i \leq n-1, i+1 \leq j \leq n}$ such that $c_{ij} \in S_{ij}$ for all $1 \leq i \leq n-1$
       and all $i+1 \leq j \leq n$ and $\sum_{i=1}^{n-1} \sum_{j=i+1}^{n} |f(c_{ij}) - f(a_{ij})| = d$ **do**
4:        $C \leftarrow$ Expand$((c_{ij})_{1 \leq i \leq n-1, i+1 \leq j \leq n});$
5:        **if** $CR(C) < 0.1$ **then return** $C$; **end if**
6:     **end for**
7:  **end for**
8:  **return** "no solution";

Function Expand$((c_{ij})_{1 \leq i \leq n-1, i+1 \leq j \leq n})$

**Input:** A candidate $(c_{ij})_{1 \leq i \leq n-1, i+1 \leq j \leq n}.$

**Output:** A PCM expanded from the candidate.

1:  **for** $i$ from 1 to $n$ **do**
2:     $c_{ii} \leftarrow 1;$
3:     **for** $j$ from 1 to $i-1$ **do** $c_{ij} = \frac{1}{c_{ji}};$ **end for**
4:  **end for**
5:  **return** $(c_{ij})_{n \times n};$

**Fig. 3.** A level-wise algorithm for solving the adjustment problem

$S_{ji} = \{1/t \mid t \in S_{ij}\}$, which can easily be guaranteed by users. The following constraints are added to the adjustment problem as well.

$$x_{ij} \in S_{ij} \qquad \text{for } i = 1, ..., n-1 \text{ and for } j = i+1, ..., n \qquad (11)$$

To summarize, the adjustment problem is expressed as the optimization problem composed of the object function (7) and constraints (8)–(11). This problem can be treated as a search problem, which finds a candidate in the candidate space $\{(c_{ij})_{1 \leq i \leq n-1, i+1 \leq j \leq n} \mid c_{ij} \in S_{ij}, 1 \leq i \leq n-1, i+1 \leq j \leq n\}$ to minimize (7) while keeping (8)–(10) hold. We stratify the candidate space according to different distances to $(a_{ij})_{1 \leq i \leq n-1, i+1 \leq j \leq n}$ and propose a level-wise algorithm for finding an optimal candidate, where level $d$ consists of all candidates $(c_{ij})_{1 \leq i \leq n-1, i+1 \leq j \leq n}$ such that $\sum_{i=1}^{n-1} \sum_{j=i+1}^{n} |f(c_{ij}) - f(a_{ij})| = d.$

The level-wise algorithm is shown in Fig. 3. Initially, the maximum distance $\sum_{i=1}^{n-1} \sum_{j=i+1}^{n} |f(c_{ij}) - f(a_{ij})|$ for all candidates $(c_{ij})_{1 \leq i \leq n-1, i+1 \leq j \leq n}$ in the candidate space is computed (line 1). The number of levels is not greater than this maximum distance. Then, each level is traversed one by one from lower ones to higher ones (lines 2–7). In level $d$, each candidate $(c_{ij})_{1 \leq i \leq n-1, i+1 \leq j \leq n}$ in the candidate space such that $\sum_{i=1}^{n-1} \sum_{j=i+1}^{n} |f(c_{ij}) - f(a_{ij})| = d$ is expanded to a PCM $C$ (line 4) and then $C$ is checked (line 5). If $CR(C) < 0.1$, then $C$ is a solution that minimizes (7) and satisfies (8)–(11), thus it is returned. Otherwise,

the processing is continued until all levels are traversed. If there is no PCM returned previously, the message "no solution" is returned (line 8). The soundness
and completeness of the algorithm, formalized below, directly follows from the
fact that all candidates in the candidate space are processed level by level in the
algorithm while the candidate space contains all solutions (note that the given
PCM $(a_{ij})_{n \times n}$ is insufficiently consistent and must not be a solution).

**Theorem 1.** *The algorithm* SearchOptimalPCM$((a_{ij})_{n \times n}, \{S_{ij} \mid 1 \le i \le n - 1, i+1 \le j \le n\})$ *returns a solution to the optimization problem composed of the object function (7) and constraints (8)–(11) if there is at least one solution, or returns "no solution" otherwise.*

Consider the complexity of the proposed algorithm. Since the execution time
is dominated by the CR computation in line 5 while each candidate needs the
CR computation exactly once, we use the number of candidates to estimate
the complexity. For a candidate in level $d$, the number of its judgements that
differ from the corresponding judgements in the original PCM is between one to
$\min\{d, \frac{(n-1)n}{2}\}$; the distance $d$ is distributed to these different judgements; each
different judgement can be smaller or larger than the corresponding judgement
in the original PCM. Hence, the number of candidates in level $d$ is up to

$$\sum_{i=1}^{\min\{d, \frac{(n-1)n}{2}\}} 2^i \binom{\frac{(n-1)n}{2}}{i} \binom{d-1}{i-1}.$$

Roughly speaking, the number of candidates in level $d$ is $O(n^{\min\{2d, (n-1)n\}})$.

## 4    An Approximation of the Proposed Method

By the complexity result, the level-wise algorithm given in Fig. 3 may not work
for cases where the resulting PCM locates at a high level. To make the proposed
method more practical, we propose an approximate algorithm for it. This algorithm, shown in Fig. 4, adopts the local search paradigm [15]; i.e., it conducts the
search in a given number of iterations, where in each iteration it looks for better
candidates in the neighborhood of the best candidate found in the previous iteration. Local search has shown its general applicability and its high efficiency in
finding local-optimal solutions to optimization problems [15]. These advantages
motivate us to apply local search to approximate the proposed method.

In our approximate algorithm (see Fig. 4), $C^* = (c^*_{ij})_{n \times n}$ is the best PCM
in the current iteration, $C' = (c'_{ij})_{n \times n}$ is the best PCM in the previous iteration, and the function Expand is the same one defined in Fig. 3. The neigh-
borhood of a candidate $(c_{ij})_{1 \le i \le n-1, i+1 \le j \le n}$ is defined as the set of candidates
$(c'_{ij})_{1 \le i \le n-1, i+1 \le j \le n}$ in the candidate space such that $\sum_{i=1}^{n-1} \sum_{j=i+1}^{n} |f(c'_{ij}) - f(c_{ij})| = 1$. In each iteration, the algorithm handles every candidate in the
neighborhood of the best candidate in the previous iteration (lines 4–12). Ev-
ery candidate is expanded to a PCM and then the CR value of the PCM is

**Algorithm 2.** LocalSearchPCM($m$, $(a_{ij})_{n \times n}$, $\{S_{ij} \mid 1 \leq i \leq n-1, i+1 \leq j \leq n\}$)

**Input:** The maximum number of iterations $m$, a reciprocal and insufficiently consistent PCM $(a_{ij})_{n \times n}$, as well as a collection of sets of allowed numbers $\{S_{ij} \mid 1 \leq i \leq n-1, i+1 \leq j \leq n\}$ for judgements in an adjusted PCM.

**Output:** An adjusted PCM or the message "no solution".

1:    $C^* \leftarrow (a_{ij})_{n \times n}$; $r^* \leftarrow CR(C^*)$; found $\leftarrow$ false;

2:    **for** $t$ from 1 to $m$ **do**

3:       $C' \leftarrow C^*$;

4:       **for** each candidate $(c_{ij})_{1 \leq i \leq n-1, i+1 \leq j \leq n}$ such that $c_{ij} \in S_{ij}$ for all $1 \leq i \leq n-1$ and all $i+1 \leq j \leq n$ and $\sum_{i=1}^{n-1} \sum_{j=i+1}^{n} |f(c_{ij}) - f(c'_{ij})| = 1$ **do**

5:         $C \leftarrow$ Expand$((c_{ij})_{1 \leq i \leq n-1, i+1 \leq j \leq n})$; $r \leftarrow CR(C)$;

6:         **if** found **then**

7:           **if** $r < 0.1$ and $\sum_{i=1}^{n-1} \sum_{j=i+1}^{n} |f(c_{ij}) - f(a_{ij})| < \sum_{i=1}^{n-1} \sum_{j=i+1}^{n} |f(c^*_{ij}) - f(a_{ij})|$ **then** $C^* \leftarrow C$; **end if**

8:         **else if** $r < r^*$ **then**

9:           $C^* \leftarrow C$; $r^* \leftarrow r$;

10:          **if** $r < 0.1$ **then** found $\leftarrow$ true; **end if**

11:         **end if**

12:       **end for**

13:       **if** $C^* = C'$ **then break**; **end if**

14: **end for**

15: **if** found **then return** $C^*$; **else return** "no solution"; **end if**

Fig. 4. An approximate algorithm for solving the adjustment problem

computed (line 5). Afterwards, the PCM is treated differentially according to different values of the flag "found". In case "found" is false, which indicates the first phase, the PCM is treated as the best PCM in the current iteration if it has a smaller CR value (line 9). If the CR value is less than 0.1, the PCM will be sufficiently consistent, and the algorithm sets "found" as true and enters into the second phase (line 10). In case "found" is true, which indicates the second phase, the PCM is treated as the best PCM in the current iteration if its CR value is less than 0.1 and it has a smaller distance to the original PCM (line 7). Note that the second phase is needed because candidates explored in later iterations may have a smaller distance than those visited in former iterations. For example, there can be a sequence of candidates like $(1, 1, 1) \rightarrow (2, 1, 1) \rightarrow (2, 2, 1) \rightarrow (2, 2, 2) \rightarrow (1, 2, 2)$ in the first four iterations. It can be seen that $(1, 2, 2)$ has a smaller distance to $(1, 1, 1)$ than $(2, 2, 2)$ does, but it is not in the neighborhood of $(1, 1, 1)/(2, 1, 1)/(2, 2, 1)$ and will not be visited in the first three iterations. The algorithm prepares to terminate if either it reaches a local optimality, i.e., it cannot find better PCMs in the current iteration (line 13), or it finishes all iterations. If the algorithm is in the second phase, which means that the best PCM so far is sufficiently consistent, it returns this PCM, otherwise it returns the message "no solution" (line 15).

The PCM returned by the approximate algorithm is reciprocal and sufficiently consistent. It is also an exact solution when the distance between its upper triangular part and the corresponding part in the original PCM is no more than two, because the whole neighborhood of the initial candidate (which corresponds to the original PCM) has been explored in the algorithm. These properties are formalized in the following theorem.

**Theorem 2.** *If the algorithm* LocalSearchPCM($m$, $(a_{ij})_{n \times n}$, $\{S_{ij} \mid 1 \leq i \leq n-1, i+1 \leq j \leq n\}$) *does not return "no solution", it must return a reciprocal and sufficiently consistent PCM* $(c_{ij})_{n \times n}$*, which is a solution to the optimization problem composed of the object function (7) and constraints (8)–(11) when* $\sum_{i=1}^{n-1} \sum_{j=i+1}^{n} |f(c_{ij}) - f(a_{ij})| \leq 2$.

We also use the number of candidates to estimate the complexity of the approximate algorithm, since the execution time is dominated by the CR computation in line 5 while each candidate needs exactly once such a computation. There are up to $(n-1)n$ candidates in the neighborhood of a candidate, thus the total number of candidates handled by the algorithm is up to $(n-1)nm$.

## 5  Experimental Evaluation

We implemented both the level-wise (exact) algorithm (see Fig. 3) and the approximate algorithm (see Fig. 4) in Java. To compare these two algorithms, we randomly generated reciprocal but insufficiently inconsistent PCMs of order 3, 4 or 5. For each order, we randomly generated 100 PCMs and then for each generated PCM, we compared both algorithms on the execution time and the distance between the adjusted PCM and the original PCM. A time limit of ten minutes was set for both algorithms to adjust a test PCM. All judgements in an adjusted PCM are restricted to numbers in the Satty scale for both algorithms. The maximum number of iterations in the approximate algorithm is set as 100. All experiments were conducted on a PC with Pentium Dual Core 2.80GHz CPU and 16GB RAM, running Win 7 (64 bit).

The average execution time for adjusting a test PCM of order 3 (resp. 4 or 5) is shown in Fig. 5 (a) (resp. (c) or (e)). The curves in the figure are drawn against the levels at which the exact solutions (i.e. adjusted PCMs) found by the level-wise algorithm locate. We simply call these levels *exact levels*. The approximate algorithm succeeds in finding a local-optimal solution in a few milliseconds for every test PCM. There are some test PCMs of order 5 for which the level-wise algorithm fails to find an exact solution within the time limit. For these test PCMs, the average execution time of the approximate algorithm is shown above the mark "19+" in the horizontal axis. The mark "19+" does not mean that the exact levels are at least 19 but only indicates the cases where the level-wise algorithm fails. It can be seen that, when the exact level increases, the execution time of the level-wise algorithm increases sharply while the execution time of the approximate algorithm increases slightly. Moreover, when the order increases from 3 to 5, the execution time of the level-wise algorithm also increases radically

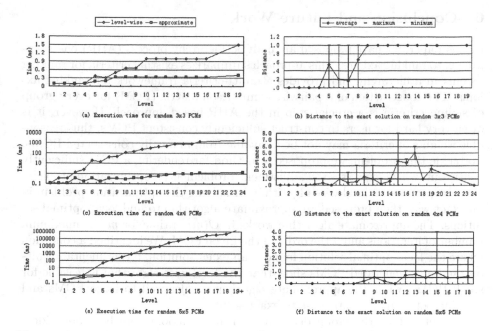

**Fig. 5.** The comparison results between the level-wise algorithm and the approximate algorithm for adjusting a randomly generated PCM

while the execution time of the approximate algorithm increases a little only. These results show that, although the level-wise algorithm is feasible when the exact level is low, it is hard to scale to large PCMs; in contrast, the approximate algorithm is highly efficient and scales well to large PCMs.

For a test PCM $(a_{ij})_{n \times n}$ of order $n$, let $(a^*_{ij})_{n \times n}$ be the exact solution found by the level-wise algorithm and $(a'_{ij})_{n \times n}$ be the solution found by the approximate algorithm. We refer to

$$\sum_{i=1}^{n-1} \sum_{j=i+1}^{n} |f(a'_{ij}) - f(a_{ij})| - \sum_{i=1}^{n-1} \sum_{j=i+1}^{n} |f(a^*_{ij}) - f(a_{ij})|$$

as the *distance* between $(a'_{ij})_{n \times n}$ and $(a^*_{ij})_{n \times n}$, simply the distance to the exact solution. The average/maximum/minimum distance for adjusting a test PCM of order 3 (resp. 4 or 5) is shown in Fig. 5 (b) (resp. (d) or (f)). The sticks and curves in the figure are also drawn against the exact levels. For the cases where the level-wise algorithm fails, the exact levels cannot be computed, thus the corresponding distances are not shown. It can be seen that, for test PCMs of order 3, the distances are at most 1, while for test PCMs of higher orders, the distances can still be small when the exact level increases. For example, for a test PCM of order 4 whose exact level is 24, the distance is 0, while for several test PCMs of order 5 whose exact levels are up to 18, the distances are also 0. These results show that the approximate algorithm can find solutions that are close to exact solutions.

# 6   Conclusion and Future Work

In this paper, we have proposed an analytic hierarchy process (AHP) based approach to ranking axioms. This approach summarizes existing criteria for ranking axioms in a hierarchy and provides a multi-criteria framework for ranking axioms. The construction of a pairwise comparison matrix (PCM) for every group of sibling criteria is a crucial step in the AHP based approach. However, it is often very hard for users to construct sufficiently consistent PCMs, thus reasonable weights of criteria may not be derivable from manually constructed PCMs. To tackle this problem, we proposed a method for rendering a PCM sufficiently consistent, which expresses the adjustment problem as an optimization problem. We first proposed a level-wise algorithm for solving the optimization problem in an exact way, then proposed an approximate algorithm to find local-optimal solutions. The approximate algorithm works in $O(mn^2)$ time for $m$ the maximum number of iterations and $n$ the order of the given PCM. We identified a condition where the local-optimal solution found is an exact solution. Experimental results on randomly generated PCMs show that the level-wise algorithm is feasible for small PCMs, while the approximate algorithm scales well to large PCMs and can find solutions that are close to exact solutions.

We notice that the group decision for ranking axioms can be more widely applicable than individual decisions. There exist two common ways to aggregate individual decisions on pairwise comparisons, namely aggregating individual judgements (AIJ) and aggregating individual priorities/weights (AIP) [9]. In the future work, we will study the methods for aggregating individual decisions on pairwise comparisons, by using the ways of AIJ and AIP, respectively.

**Acknowledgements.** This work is partly supported by the NSFC under grants 61005043, 61375056 and 71271061, the Undergraduate Innovative Experiment Project in Guangdong University of Foreign Studies, as well as the Business Intelligence Key Team of Guangdong University of Foreign Studies (TD1202).

# References

1. Aguarón, J., Moreno-Jiménez, J.M.: The geometric consistency index: Approximated thresholds. European Journal of Operational Research 147(1), 137–145 (2003)
2. Barzilai, J.J.: Deriving weights from pairwise comparison matrices. Journal of the Operational Research Society 48, 1226–1232 (1997)
3. Cao, D., Leung, L.C., Law, J.S.: Modifying inconsistent comparison matrix in analytic hierarchy process: A heuristic approach. Decision Support Systems 44(4), 944–953 (2008)
4. Deng, X., Haarslev, V., Shiri, N.: Measuring inconsistencies in ontologies. In: Franconi, E., Kifer, M., May, W. (eds.) ESWC 2007. LNCS, vol. 4519, pp. 326–340. Springer, Heidelberg (2007)
5. Du, J., Qi, G., Shen, Y.-D.: Lexicographical inference over inconsistent DL-based ontologies. In: Calvanese, D., Lausen, G. (eds.) RR 2008. LNCS, vol. 5341, pp. 58–73. Springer, Heidelberg (2008)

6. Du, J., Qi, G., Shen, Y.: Weight-based consistent query answering over inconsistent SHIQ knowledge bases. Knowledge and Information Systems 34(2), 335–371 (2013)
7. Du, J., Shen, Y.: Computing minimum cost diagnoses to repair populated DL-based ontologies. In: Proceedings of the 17th International World Wide Web Conference (WWW), pp. 575–584 (2008)
8. Ergu, D., Kou, G., Peng, Y., Shi, Y.: A simple method to improve the consistency ratio of the pair-wise comparison matrix in anp. European Journal of Operational Research 213(1), 246–259 (2011)
9. Forman, E., Peniwati, K.: Aggregating individual judgments and priorities with the analytic hierarchy process. European Journal of Operational Research 108, 165–169 (1998)
10. Huang, Z., van Harmelen, F., ten Teije, A.: Reasoning with inconsistent ontologies. In: Proceedings of the 19th International Joint Conference on Artificial Intelligence (IJCAI), pp. 454–459 (2005)
11. Hunter, A., Konieczny, S.: Measuring inconsistency through minimal inconsistent sets. In: Proceedings of the 11th International Conference on Principles of Knowledge Representation and Reasoning (KR), pp. 358–366 (2008)
12. Kalyanpur, A., Parsia, B., Sirin, E., Cuenca-Grau, B.: Repairing unsatisfiable concepts in OWL ontologies. In: Sure, Y., Domingue, J. (eds.) ESWC 2006. LNCS, vol. 4011, pp. 170–184. Springer, Heidelberg (2006)
13. Lam, S.C., Pan, J.Z., Sleeman, D.H., Vasconcelos, W.W.: A fine-grained approach to resolving unsatisfiable ontologies. In: Proceedings of the International Conference on Web Intelligence (WI), pp. 428–434 (2006)
14. Li, H., Ma, L.: Detecting and adjusting ordinal and cardinal inconsistencies through a graphical and optimal approach in ahp models. Computers & OR 34(3), 780–798 (2007)
15. Michiels, W., Aarts, E., Korst, J.: Theoretical aspects of local search. Springer (2007)
16. Mu, K., Liu, W., Jin, Z.: A general framework for measuring inconsistency through minimal inconsistent sets. Knowledge and Information Systems 27(1), 85–114 (2011)
17. Qi, G., Hunter, A.: Measuring incoherence in description logic-based ontologies. In: Aberer, K., et al. (eds.) ISWC/ASWC 2007. LNCS, vol. 4825, pp. 381–394. Springer, Heidelberg (2007)
18. Qi, G., Ji, Q., Pan, J.Z., Du, J.: Extending description logics with uncertainty reasoning in possibilistic logic. International Journal of Intelligent Systems 26(4), 353–381 (2011)
19. Saaty, T.L.: The analytic hierarchy process: planning, priority setting, resource allocation. McGraw-Hill, New York (1980)
20. Vaidya, O.S., Kumar, S.: Analytic hierarchy process: An overview of applications. European Journal of Operational Research 169(1), 1–29 (2006)
21. Xu, Z., Wei, C.: A consistency improving method in the analytic hierarchy process. European Journal of Operational Research 116(2), 443–449 (1999)

# Learning Domain-Specific Ontologies from the Web

Wenkai Mo[2], Peng Wang[1,2], Haiyue Song[2], Jianyu Zhao[2], and Xiang Zhang[1,2]

[1] School of Computer Science and Engineering, Southeast University, Nanjing, China
[2] College of Software Engineering, Southeast University, Nanjing, China
pwang@seu.edu.cn

**Abstract.** This paper proposes an approach of learning domain-specific ontologies from the Web. First, a webpage is segmented into text blocks by analyzing the visual features and DOM structures. Second, text blocks will be labeled by Conditional Random Fields (CRFs) model. Third, a local ontology of the webpage is constructed based on vision tree and labeled text blocks. Finally, the ontology for a website is generated by merging the local ontologies. Our experimental results on real world datasets show that the proposed method is effective and efficient for domain-specific ontology learning, and the results have average 0.91 F-measure for concepts, instances and *subclass-of* relations.

**Keywords:** ontology, ontology learning, condition random fields, page segmentation.

## 1 Introduction

Discovering knowledge from the Web is an open problem because the Web information are semi-structured, heterogeneous, and in various representations. Ontology, a kind of knowledge representation, can be used to describe the knowledge in the Web. It is also the basic semantic data for the Semantic Web and Linked Data applications. However, due to the huge Web information and the rapid evolution of Web contents, manually constructing ontologies is impossible. Therefore, finding automatic or semi-automatic ontology learning methods from the Web is very necessary.

Learning ontology from free text and semi-structured data has been studied in past decades [1]. Not only natural language processing techniques such as part-of-speech tagging, syntactic analysis, semantics analysis and name-entity recognition, but also machine learning techniques and rules-based methods are used for automatic ontology generation. This paper focuses on the problem of learning ontology from webpages, which is the popular semi-structured data. Currently, most webpages are in Hyper Text Markup Language (HTML), and with Cascading Style Sheets (CSS) and Java-Script for auxiliary. HTML pages can be parsed to Document Object Model (DOM) trees. Therefore, text nodes of DOM tree can be treated as concepts or instances or properties or property values, and hierarchical relationship between concepts can be learned from the DOM tree structure [2]. However, many webpages do not obey W3C standard, parsing these webpages for ontology learning may lead to errors. Another way for learning ontology property is based on knowledge base [3], such as World Net. However, new emerging concepts or instances cannot be found using these knowledge bases.

G. Qi et al. (Eds.): CSWS 2013, CCIS 406, pp. 132–146, 2013.
© Springer-Verlag Berlin Heidelberg 2013

This paper proposes an approach of learning domain-specific ontology from the Web. We demonstrate our approach on the academic conference domain, but we must emphasis that our approach is a generic ontology learning approach. For each web-page in a website, concepts and instances are learned by machine learning techniques, and relationship learning is based on the webpage vision tree. Reference links between webpages are used to merge local ontologies to an ontology for the website.

The rest of paper is organized as follows. Related works are reviewed in Section 2. Overview of our approach is in Section 3. Page segmentation algorithm is discussed in Section 4. CRFs for text blocks classification is described in Section 5. Constructing webpage local ontology is given in Section 6. Merging local ontology to the ontology of the website is given in section 7. Evaluation results are reported in Section 8. Finally, concluding remarks are given in Section 9.

## 2    Related Work

Many approaches for extracting ontology automatically have been proposed. However, only few works focused on learning ontologies from the Web. Du et al. proposed a six-phase process to extract ontology from HTML websites based on the DOM tree, the words' TF-IDF in the webpage and HTML links [2]. However, webpage designers will emphasize the important key words with some special visual information, such as the size of fonts and the weight of words. Considering visual features, the words with low TF-IDF weight may be also very important. Arasu and Hector proposed a method of extracting structured data from webpages [4]. It used sets of words that have similar occurrence patterns in webpages to construct the webpages templates, which described how data was encoded into webpages. However, this method could only deal with some webpages having similar templates. A very important point is that both Du et al. and Arasu did not consider visual features. Visual features reflect human-beings' reading habit. The items with different visual features would attract people's attention, and they also contains important information.

Cai et al. proposed a method to split the webpage into some sections based on visual features called Vision-based Page Segmentation (VIPS) algorithm [5]. This algorithm first extracted all the suitable blocks from the HTML DOM tree, then tried to find the separators between these extracted blocks. The algorithm did this process recursively, until it could not find separators. When the algorithm found the separators, it considered the page layout features. Wang et al. used VIPS algorithm to segment webpage to some sections and remove the navigation section of the webpage and labeled the rest of sections [6]. However, only with these labels we cannot transfer them into ontology. Zhu et al. proposed a method to understand webpages [7, 8]. He used a vision tree to represent the webpage. Then CRFs model was used to label this vision tree. Both Peng and Zhu have labeled the vision tree. However, they didn't extract ontology based on these labels. Yao has proposed a method to extract ontology in the Researcher Profiling [9]. They first defined the schema of the researcher profile by extending the FOAF ontology [10], and then used CRFs model to learn the instance of the concept in predefine schema. But it just learned the entity in the schema they predefined, some new concepts in the webpage could not be learned. The method could solve the profile webpages, but not for other kinds of webpages in the Web.

# 3    Overview of the Approach

For convenience of discussion, this paper chooses academic conference websites as the research object. Our method is shown as Fig. 1. For a specific website $WS$, it consists many webpages $WP_1$, $WP_2$, ... $WP_n$, namely, $WS = (WP_1, WP_2, ... WP_n)$. For each $WP_i$ in $WS$, first we use a page segmentation algorithm to obtain vision tree of $WP_i$. Then we prune the vision tree by removing useless items such as pictures, navigation of $WP_i$. After that, we traverse the vision tree in deep first search (DFS), and then get a sequence of text blocks. We use CRFs model to label the sequence of text blocks. After the labeling, we use vision tree, labels and very little domain knowledge to construct local ontology $O_i$ of $WP_i$. For a $WS$, we can get a set of local ontology $O = (O_1, O_2, ... O_n)$ for each $WP_i$. Finally, we merge these local ontologies to get $WS$ ontology.

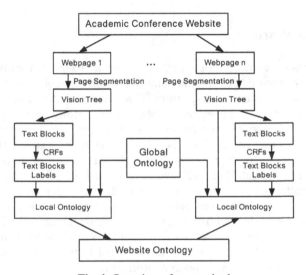

**Fig. 1.** Overview of our method

# 4    Page Segmentation Algorithm

Intuitively, people regard a webpage as combination of several semantic objects, rather than a single object. Users usually expect specific function blocks such as a navigation bar and an advertising area will appear at a fixed position. In fact, through surveying a large amount of conference webpages, we find that although different conference websites adopt different design styles and layouts, most conference webpages are composed of some stationary function blocks. Therefore, we first apply vision-based page segmentation (VIPS) algorithm [5] to the conference webpages.

The VIPS algorithm is based on the page layout features, and it first extracts all the suitable blocks from the HTML DOM tree, then tries to find the separators between these extracted blocks. Here, separators denote the horizontal or vertical lines in a webpage that visually cross with no blocks. Finally, based on these separators, the semantic structure for the webpage is constructed. VIPS algorithm employs a top-down approach, which is very effective.

Fig. 2 shows an example of vision-based content structure for *Call for Papers* page of VLDB2011 conference. It illustrates the layout structure and the vision-based content structure of the page. In the first level, the original webpage has three objects or visual blocks VB1~VB3 and two separators $\varphi^1\sim\varphi^2$. Then we can further construct sub content structure for each sub webpage. For example, VB2 has two sub-objects VB2_1~VB2_2 and one separator $\varphi_2^1$. It can be further analyzed recursively.

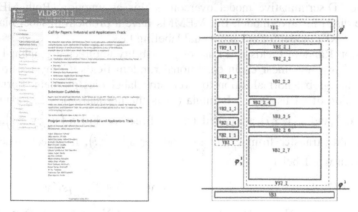

**Fig. 2.** The layout structure and vision-based content structure of a webpage

VIPS can obtain good segmentation results for most webpages, but we find it will lose important information when deal with some webpages. For example, important submission deadline information would be lost. Through the intensive analysis, we find such result is caused by the reason: VIPS algorithm is based on vision features of the page elements, so the display of segmented adjacent semantic blocks is the same (in this case, they are in bold), while the lost part is not in bold, so VIPS ignores the blocks whose display is inconsistent. Therefore, we introduce DOM-based analysis to improve the VIPS segmentation results [6].

## 5    Text Block Classification

After getting text blocks by the page segmentation, we will assign a label to each text block based on CRFs model for ontology construction. Because CRFs is a kind of sequential labeling model, we traverse the vision tree use DFS algorithm to get a sequence of text blocks and represent each text block to feature vector.

### 5.1    Condition Random Fields

We use classification model to predict text block categories based on input text block's feature vectors. There are many famous existing classification algorithms in machine learning such as C4.5 [12], Bayes Network[13] and Support Vector Machine (SVM) [14]. However, in a webpage, texts with high similarity will aggregate together, especially the date, co-chair and research topic information. Consider the aggregation, sequential labeling model may do better than classification model.

There are many famous sequential labeling models such as Hidden Markov Model (HMM) [15], Maximum Entropy Markov Models (MEMMs) [16] and Conditional Random Fields (CRFs) [11]. HMM is a kind of generative model. For generative model, it defines a joint probability distribution p($X$, $Y$) over observation and label sequences. However, strong independence assumption must be considered in generative model. Discriminative model overcomes this assumption. Both MEMMs and CRFs are discriminative model. But MEMMs have a problem of label bias [11]. So we choose CRFs model for our sequential labeling task.

Condition Random Fields is a kind of probability undirected graphical model. In our method, we use linear chain conditional random fields. Compared to MEMMs, CRFs do the global normalization to overcome label bias problem. Formula (1) is basic equation of CRFs model. Formula (2) is the normalization factor in (1). Both $t_k\left(y_{i-1}, y_i, x, i\right)$ and $s_l(y_i, x, i)$ are called feature function. For $t_k\left(y_{i-1}, y_i, x, i\right)$, it reflects the probability transfer from $y_{i-1}$ to $y_i$. And for $s_l(y_i, x, i)$, it reflects discriminative process from $x$ to label $y_i$.

$$P(Y|X) = \frac{1}{Z(X)}\exp\left(\sum_{i,k}\lambda_k t_k(y_{i-1}, y_i, x, i) + \sum_{i,l}\mu_l s_l(y_i, x, i)\right) \tag{1}$$

$$Z(X) = \sum_y \exp\left(\sum_{i,k}\lambda_k t_k(y_{i-1}, y_i, x, i) + \sum_{i,l}\mu_l s_l(y_i, x, i)\right) \tag{2}$$

In our method, $X$ is sequence of text blocks of a webpage, $Y$ is sequence of labels, each $y_i$ in $Y$ must satisfy $y_i \in L$, where $L$ will be defined later. So we must evaluate the label sequence $Y^*$ for input sequence X which satisfies equation (3). We use Viterbi Algorithm [17] to evaluate $Y^*$.

$$Y^* = \text{argmax}_Y\, P_w(Y|X) \tag{3}$$

Before evaluating $Y^*$, we must estimate parameters $\lambda_k$ and $\mu_l$ for each feature function. We maximize log likelihood of the training data to estimate parameters. The log likelihood is equation (4). We use L-BFGS Algorithm [18] to estimate parameters.

$$L(w) = \sum_{j=1}^{N}\left(\sum_{i,k}\lambda_k\, t_k\left(y_{i-1}, y_i, x, i\right) + \sum_{i,l}\mu_l s_l\left(y_i, x, i\right)\right) - \sum_{j=1}^{N}\log Z_w\left(x_j\right) \tag{4}$$

## 5.2    Categories of Text Blocks

Ontology can be defined as $O=(C, R, I)$, where $C$ is a set of concept, $I$ is a set of instance, and $R$ is a set of relationship or property. We just learn hierarchical relationship between concepts and learn instances for corresponding concept. Global ontology definition is based on domain-specific knowledge which we are interested in. For academic conference webpages, ontology concepts are *Conference*, *Date*, *Research Topic* and *Person*.

In order to construct local ontology for a webpage, the category of a text block should be learned first. Based on the global ontology, we define label set $L$ in CRFs model.

$$L = \{Conc, Conf, Date, Per, RT, Ins, NR\}$$

- *Conc*: If a text block is category *Conc*, then the text is a concept in ontology, no matter what kind of concept it is.
- *Conf*: If a text block is category *Conf*, then the text is an instance of *Conference* in our global ontology. It contains information of when and where the conference will hold. This category always appears in the top of a webpage.
- *Per* and *RT*: If a text block is category *Per* or *RT*, then the text is an instance of a person concept or a research topic concept.
- *Date*: If a text block is category Date, then the text is an instance of a date concept. Or the text is a pairwise of *(Date_Concept ,Date)*. Because sometimes page segmentation algorithm cannot segment the pairwise to two text blocks as Fig. 3(a) shows but sometimes it can be split to two blocks as Fig. 3(b) shows.
- *Ins*: If a text block is category *Ins*, then the text is an instance of a new concept we didn't predefine.
- *NR*: If a text block is category *NR*, then the text block is neither a concept nor an instance. This kind of text block is often very long. Maybe it contains some knowledge, but we cannot find them directly unless using some free text ontology extracting technologies to deal with these text blocks.

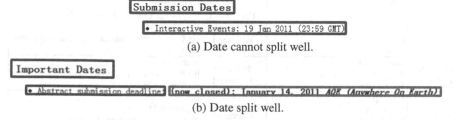

(a) Date cannot split well.

(b) Date split well.

**Fig. 3.** Example of VIPs Algorithm cannot work well in split Date

## 5.3    Feature Selection

We must select features and use feature vectors to represent each text block to build CRFs model. We select two kinds of features as Table 1 shows.

- **Visual Feature.** Intuitively, the more important the text is, the more emphasis would be laid on the text. Visual features include text display style, structure information in webpage, structure information in text block, and location in the webpage. Text display style consists of font size and font weight. Structure information in the webpage depends on whether it is in tag <h> or whether it is in a list, while structure information in text block has much to do with several factors, i.e. the number of words in text, frequency of capital words in text and whether the text ends with a colon. Location of a text block in the webpage is also important. We divide a webpage into three parts: top part, middle part and tail part. Location feature reflects which part the text block is in.
- **Sematic Feature.** Sematic feature mainly reflects whether a text block contains hyperlink or whether it contains special keywords such as country names, months and institutions.

Table 1. Visual Features and Semantic Features

| Visual Features | Description | Semantic Features | Description |
|---|---|---|---|
| Font Size | value: {-1,0,1}<br>fs: Font size appears most frequently in the page<br>0: Font size is equal to fs<br>-1: Font size is smaller than fs<br>1: Font size is larger than fs | ContainUrl | value:{0,1}<br>0: Not contain Url<br>1: Contain Url |
| Font Weight | value:{0,1}<br>0: Not in bold<br>1: In bold | Contain Month | value:{0,1}<br>0: Not contain month<br>1: Contain month |
| Start With H | value:{0,1}<br>0: Not start with tag <h><br>1: Start with tag <h> | Topic Keyword | value:{0,1}<br>0: Not contain Concept Keyword<br>1: Contain Concept Keyword |
| Start With Li | value:{0,1}<br>0: Not start with tag <li><br>1:  Start with tag <li> | Concept Keyword | value:{0,1}<br>0: Not contain Concept Keyword<br>1: Contain Concept Keyword |
| Number Of Words | value:{0,1,2}<br>Number or words in:<br>0: (0,15]<br>1: (15,30]<br>2: (30,+∞) | PreTopic Concept Keyword | value:{0,1}<br>0: No RT Concept Keyword in previous text block<br>1: Contain RT Concept Keyword in previous text block |
| Frequency Of Capital Words | value:{0,1,2}<br>Frequency Of Capital Words in:<br>0: (0.8,1)<br>1: (0.5,0.8]<br>2: (0,0.5] | PrePerson Concept Keyword | value:{0,1}<br>0: No Per Concept Keyword  in previous text block<br>1: Contain Per Concept Keyword in previous text block |
| End With Colon | value:{0,1}<br>0: Not end with clone<br>1: End with clone | Institute Keyword | value:{0,1}<br>0: Not contain Institute Keyword<br>1: Contain Institute Keyword |
| Location In Webpage | value:{0,1,2}<br>Located in:<br>0: Title Part<br>1: Main Text Part<br>2: Not Related Part | Cohuntry Keyword | value:{0,1}<br>0: Not contain Country Keyword<br>1: Contain Country Keyword |

An interesting thing is that using CRFs model to label may cause an error accumulation problem in some specific scenes. If the first item in a list is predicted wrong, because of the probability transfer function $t_k \left( y_{i-1}, y_i, x, i \right)$, it will cause text blocks in the same list are all predicted wrong. So we use features "PreTopicConcept Keyword" and "PrePersonConcept Keyword" to improve the prediction accuracy of first text block in the list.

# 6     Constructing Webpage Ontology

Page segmentation algorithm constructs vision tree by finding the separators between vision blocks recursively. If there are several separators in a block, then the algorithm will split the block as Fig. 4 (a) shows. Finally, in vision tree, information is stored separately in leaf nodes and inner nodes reflect structure information among their children. Children of an inner node may have some similarities in visions or semantics.

(a) Process of VIPS Algorithm          (b) Split *Date* to *DATE_CONC* and *Date*

**Fig. 4.** Process in vision tree

After processing of CRFs, each text block has already been labeled. To construct ontology, the first thing we need to do is to clean text blocks with label *NR* and *Conf*. As is defined above, *NR* text blocks are not related to ontology construction, so it is better to remove them directly. Text blocks with label *Conf* are instances of concept *Conference* in global ontology. We remove these nodes and then mark them to be instances of concept *Conference* in global ontology. If some virtual nodes become leaves after cleaning *NR* and *Conf* text blocks, we delete these virtual leaf nodes recursively.

Second, the relationship between text blocks should be discovered. In order to build ontology, it is necessary to find *subclass-of* relationship between concepts and *is-a* relationship between instances and corresponding concepts. In this paper, we define type $T \in \{DATE, PER, RT, NEW\}$ for corresponding instances and concepts. *DATE* is for label *Date*, *PER* is for label *Per*, *RT* for label *RT* and *NEW* is new type we didn't predefine for label *Ins*. For convenient, we use *CONC* to represent concept and use *INS* to represent labels.

$$INS \in \{RT, Ins, Date, Per\} \subseteq L$$

- **Split Date type if necessary.** *Date* labels can be an instance of a date concept or a pairwise of *(DATE_CONC,DATE)*. We use regular expressions to split pairwise if it is. It changes the structure of node in vision tree as Fig. 4 (b) shows.
- **Find *is-a* relationship.** Two kinds of structure in vision tree as Fig. 5 shows can be found from bottom to up.
  - For a local structure likes the left of Fig. 5 (a), let *INS* be type $T$, we change *CONC* to *T_CONC* for further relationship finding and change the structure to the right of (a).
  - If a local structure is like the left of Fig. 5 (b). Use a voting algorithm (Algorithm 1) to confirm which type the *CONC* is. In voting algorithm, if an instance of a type $T$ have a ratio greater than $\lambda$ (line 3), then *CONC* is *T_CONC*. If we can make sure type $T$ for *CONC*, and if text node in the same level is not instance of type $T$, it must be labeled wrong so change it to corresponding label of

(a) *is-a* Relationship situation 1          (b) *is-a* Relationship situation 2

**Fig. 5.** Find *is-a* Relationship

| Algorithm 1: Voting Algorithm in *is-a* relationship finding |
|---|
| Input: A tree, root is a concept without type and children are all INS |
|     λ for voting rate |
| Output: Type of Concept |

| | |
|---|---|
| 1 | **Begin** |
| 2 |   **foreach** i in INS **do** |
| 3 |     **if** (count(children,i)/count(children)>=λ) **then**//Compute ratio of i in children |
| 4 |       correct(children,i) //Change all children's label to label i |
| 5 |       **return** type(i) //Return the corresponding type of i |
| 6 |   **End** |
| 7 |   **return** NEW   // Cannot find a suitable type then we treat it to a new concept |
| | **End** |

type $T$ (line 4). If it is not sure what type is the $CONC$, it is a new concept, let $T=NEW$ (line 7). We set $λ=0.9$ in our experiment.

- **Find *subclass-of* relationship.** After finding *is-a* relationship, structure of vision tree will change a lot, many virtual nodes are replaced. Next we will find relationship between concepts based on the changed vision tree from bottom to up. There are two situations we must consider as follows.

  - If a local structure likes the left of Fig. 6 (a), $T\_CONC$ 2 in left of the figure may have some instances or some sub-concepts because $T\_CONC$ 2 is converted from a virtual node before. Both $CONC$ 1 and $T\_CONC$ 2 come from a same vision block, so change $CONC$ 1 to $T\_CONC$ 1. And change the left structure to right one to represent *subclass-of* relationship.
  - If a local structure likes Fig. 6 (b). This kind of situation is the most difficult one to deal with. Children of virtual node $C$ must be sub-concept or instance of concept $B$. However, we cannot make sure what type the concept $B$ is. Maybe there are some concept nodes have no type like node $E$. These kinds of concepts like "Sponsor" have no instance because instances of these concepts are pictures. We treat them as new concepts, and assign $T=NEW$ to them. Maybe there are some instance nodes like node $F$. These instances are instance of concept $B$. There are also some concept nodes with type like node $D$ and node $G$. Both instance nodes and concept nodes will affect the type of concept $B$. To deal with this problem, we can also use voting algorithm (Algorithm 2). However, in this situation, we must assign a voting weight to each node. Calculate the weight of each child of $C$ (line 3). If weight ratio of type $T$ is greater than $λ$ (line 6), then $B$ node is a $T\_CONC$. Then we must do some cleaning for the children of $C$. If $T$ is not type $NEW$ (line 9), change all $Ins$ labels, to label of type $T$ because they are wrongly classified (line 25-33). For other nodes, we move those which are not type $T$ to the corresponding location in global ontology. For weight assignment, we use an attenuation rate $α$. For example, if $D$ node in Fig. 4 (b) has 10 children of $RT$, then the weight of D is $10*α$ (line 14-24). In our experiment, we set $λ=0.9$ and $α=0.15$.

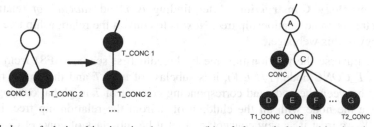

(a) *subclass-of* relationship situation 1          (b) *subclass-of* relationship situation 2

**Fig. 6.** Find *subclass-of* relationship

| Algorithm 2: Voting Algorithm in *subclass-of* relationship finding |
|---|
| Input: A tree root is a concept without type and children are mixed. |
| λ for voting rate |
| α for attenuation rate |
| Output: Type of Concept |

```
1    Begin
2      foreach child in root do
3        calcuateWeight(child,α)
4      End
5      foreach i in INS do
6        if (weight(children,i)/weight(children)>=λ) then
7          if ( i== Ins ) then return NEW
8          Else
9            cleanTheTree(root,i)
10           return type(i) //Return the corresponding type of i
11     End
12     return NEW   // Cannot find a suitable type, treat it to a new concept
13   End
14   Function calcuateWeight (node,α)
15   Begin
16     if (not hasChildren (node)) then
17         node.weight = 1; return 1
18     totalWeight = 0
19     foreach n in node do
20        totalWeight += calcuateWeight(n,α)
21     End
22     node.weight = totalWeight * α
23     return node.weight
24   End
25   Function cleanTheTree (root,i)
26   Begin
27     foreach child in root
28       if (not isConcept(child.label)) then child.label = i
29       else if (type(child.label) is not type(i)) then
30           if (child.label == INS) then child.label = i
31           else moveToGlobalOntology(child)
32     End
33   End
```

- **Local ontology Construction.** After finding *is-a* and *subclass-of* relationship, vision tree become relationship tree. Next is to convert the relationship tree to local ontology of this webpage.

  - First traverse the relationship tree by breadth first search (BFS) from root. If find $T\_CONC$ node $(T{\neq}NEW)$, it is subclass of type $T$ and then build relationship between the node and corresponding concept of $T$ in global ontology. And delete the node with all the children of it from the relationship tree. For $T = NEW$, we treat it as new concept, and let it be sub-class of concept *Conference* in global ontology. If we find $CONC$ node without type $T$, we assign $T = NEW$ to it and let it be a new concept. If we find $INS$ $(INS{\neq}Ins)$ node, this node is an instance of corresponding concept of $T$, we move it to global ontology as an instance of concept $T$. For $INS{=}Ins$, we cannot make sure it belongs to which concept, so delete it.

  - Second we correct some *subclass-of* and *is-a* relationships. For all new concepts we find in webpage, if there are some other concepts or instances of type $T$ $(T{\neq}NEW)$ in their children, we build relationship between them and corresponding concept in global ontology as Fig. 7 shows.

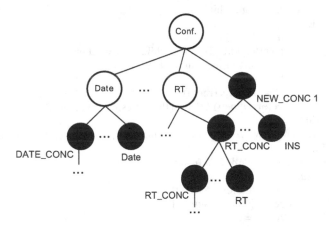

**Fig. 7.** Local ontology Construction

With the operations above completed, the vision tree transforms into a graph $G = (E,V)$. Each vertex in $G$ represents a concept or an instance, and each edge in $G$ reflects the relationship between vertexes. So $G$ is the local ontology we learn from the webpage. We use RDF triples to store the local ontology.

# 7    Merging Webpage Ontologies to the Website Ontology

We have local ontology of each webpage in a specific website. Each website has "Homepage" page, and webpages are organized by this "Homepage" page through navigation section, which has reference links to other webpages. This is clue of merging webpage ontology to website ontology.

Ontology merging contains two parts, first is merge webpage ontology to website ontology. Second we must eliminate inconsistency in this website ontology. Now, we have already got ontology $O_1, O_2, ...O_n$ for each webpage in website. Let $O_1$ be $O_{ws}$, and then merge the rest of $O_i$ one by one to $O_{ws}$ based on navigation of the website in "Homepage". During the process of merging, there may appear some duplicate concepts or instances. We use our previous work on ontology matching [19] to find them. And delete the duplication when necessary.

We can get website ontology after the merging process. Inconsistencies between concepts or instance may exist in website ontology. To eliminate inconsistencies, we use ontology debugging technology to find conflicts [20] and eliminate them manually. We can get the final website ontology after inconsistency elimination.

# 8    Experiment

## 8.1    Experiment Preparation

We implement our method in Python and Weka. The experimental results are obtained on a PC with 2.40GHz CPU, 2GB RAM and Windows 7.

Experiments contain two parts. First one is model comparison. We compare CRFs, SVM, Naïve Bayes, C4.5 models. Second one is evaluation of webpage ontology construction algorithm. During experiments, our system can process a conference webpage in average 1.03 seconds, which contains 0.18 seconds for segmenting the webpage, 0.7 seconds for classifying text blocks and 0.15 seconds for ontology construct.

We collect 20 academic conference websites that have 103 different webpages and 5472 labeled text blocks in the field of computer science for experiment. Randomly select 15 websites as training set for constructing CRFs model. The other 5 conference websites are testing set.

We use Precision, Recall and F-Measure as criteria to measure the system performance.    $Precision = \frac{A \cap B}{A}$, $Recall = \frac{A \cap B}{B}$, $F - Measure = \frac{2 \times Precision \times Recall}{Precision + Recall}$, where A is the result obtained by our system, B is the reference result built manually.

## 8.2    Model Comparison Experiment

First experiment is comparing the models for learning ontology concepts and instances, namely, labeling text blocks after page segmentation. We employ four different classification models: Decision tree, Naïve Bayesian, Support Vector Machine and Conditional Random Fields. The experimental results are given in Table 2. This experiment uses same training data set for constructing classifiers.

From the experimental results, we can see that CRFs have higher average precision, recall and F-Measure than other three models. Though the CRFs produces 0.84 precision, 0.80 recall and 0.82 F-measure, it is also not good enough. We can find that *Ins* and *NR* labels predict not well. For *Ins* label, it is the instance of a new concept. It is easy to be labeled to *RT* if it appears in a list and *NR* if it doesn't appear in a list. *NR* is also easily labeled to *Ins*. *Conf* also predict not well. *Conf* is name of a conference, it always appear in top of a webpage. It is often classified to *Conc*, like "THE TWENTY-FIFTH CONFERENCE ON ARTIFICIAL INTELLIGENCE (AAAI-11)" text block in AAAI-2011. We can even make sure whether it is an instance of a *Conf* or a *Conc* manually. Because labels will affect ontology construction, we will do some post-processing based on the vision tree to make them more accurate in the future.

**Table 2.** Model Comparison

|  | C4.5 | | | SVM | | | Naïve Bayes | | | CRFs | | |
|---|---|---|---|---|---|---|---|---|---|---|---|---|
|  | P | R | F | P | R | F | P | R | F | P | R | F |
| Conc | 0.81 | 0.89 | 0.85 | 0.74 | 0.89 | 0.81 | 0.80 | 0.79 | 0.79 | 0.86 | 0.88 | 0.86 |
| RT | 0.79 | 0.96 | 0.87 | 0.78 | 0.96 | 0.86 | 0.80 | 0.97 | 0.87 | 0.88 | 0.95 | 0.91 |
| Per | 0.94 | 0.93 | 0.94 | 0.91 | 0.93 | 0.92 | 0.90 | 0.95 | 0.93 | 0.94 | 0.97 | 0.95 |
| NR | 0.66 | 0.66 | 0.66 | 0.61 | 0.58 | 0.59 | 0.53 | 0.64 | 0.58 | 0.72 | 0.69 | 0.70 |
| Conf | 0.75 | 0.68 | 0.71 | 0.76 | 0.63 | 0.69 | 0.74 | 0.63 | 0.68 | 0.83 | 0.75 | 0.77 |
| Ins | 0.75 | 0.38 | 0.51 | 0.73 | 0.32 | 0.45 | 0.69 | 0.35 | 0.46 | 0.68 | 0.44 | 0.61 |
| Date | 0.88 | 0.95 | 0.91 | 0.88 | 0.84 | 0.86 | 0.90 | 0.88 | 0.89 | 0.94 | 0.93 | 0.93 |
| Avg. | 0.80 | 0.78 | 0.78 | 0.77 | 0.74 | 0.74 | 0.76 | 0.74 | 0.74 | **0.84** | **0.80** | **0.82** |

## 8.3    Webpage Ontology Constrtion Experiment

We choose about 20 webpages in different 6 conferences to analyze the result of ontology construction. We evalutate concept, *subclass-of* relationship and *is-a* relationship manually. Some texts like *THE TWENTY-FIFTH CONFERENCE ON ARTIFICIAL INTELLIGENCE (AAAI-11)*, we treat them as a concept are not quite wrong.

We can see from Table 3 that *is-a* relationship has a relatively good result. Although some labels must be wrong, because of structure of vision tree, they can find the corresponding concept. However, the wrong label will affect *subclass-of* finding. Such as if a list *Ins* are wrongly labeled to *RT*, the corresponding concept will be a subclass of Reseach Topic in global ontology instead of new concept. Some concepts especially new concepts sometimes cannot find their instance, because it labels their instance to *NR*. Sometimes this kind of lost concepts will accumulate in a list. Some new concepts have no instance or subcalss, so the number of concept with *subclass-of* is less than the learned concepts.

**Table 3.** Result of Construct Webpage Ontology

|  | Concept | *is-a* | *subclass-of* |
|---|---|---|---|
| Correct | 130 | 534 | 86 |
| Excessive | 18 | 11 | 16 |
| Lost | 16 | 31 | 5 |
| Precision | 89% | 98% | 84% |
| Recall | 88% | 95% | 95% |
| F-Measure | 88% | 96% | 89% |

# 9    Future Works and Conclusion

We can see find *subclass-of* and *is-a* relationship can work relatively well based on the vision tree. However, some situations in vision tree we haven't considered yet. Such as if a vision block has more than one concept and many kinds of instances at the same time. Although this kind of situation seldom appears, our method can work well in the situation. We can use the principle of proximity to match instance to corresponding concept.

After using CRFs model, many labels are wrong. Labels are important because they will affect relationship finding. We must do some post-processing work to make these labels more accurate. A good method is also using the structure of vision tree. Our method mainly deals with the most important part *subclass-of* relationship finding and *is-a* relationship finding in ontology construction. We will find general relationship in the future work.

How to merge local ontologies to website ontology is another problem. We can use semantic information such as distance between two text blocks to merge similar concept. We also must do inconsistency eliminate after we get website ontology.

This paper proposes a method to construct ontology from webpage based on vision tree and labels of the webpage. Through learning, we can get some concepts, and use vision tree to get the relationship. If a concept is subclass of concept we predefine in global ontology, we build corresponding relationship. Otherwise, the concept will be a new concept we find in the webpage. Through this way, we can expand our global ontology to webpage ontology. After getting webpage ontology, we merge them to website ontology. In this way, we extract knowledge from information in webpage for further use. As manually labeling data and building CRFs model is also relevant to other domains and costs little. The solution proposed in this paper can be applied to other domains easily.

**Acknowledgments.** This work is supported by the NSF of China (61003156 and 61003055).

# References

1. Wong, W., Liu, W., Bennamoun, M.: Ontology Learning from Text: A Look Back and into the Future. ACM Computing Surveys 44(4), 20–36 (2012)
2. Du, T.C., Li, F., King, I.: Managing knowledge on the Web - Extracting ontology from HTML Web. Decision Support Systems 47(4), 319–331 (2009)
3. Alani, H., Kim, S., Millard, D.E., et al.: Automatic Ontology-Based Knowledge Extraction from Web Documents. IEEE Intelligent Systems 18(1), 14–21 (2003)
4. Arasu, A., Garcia-Molin, H.: Extracting Structured Data from Web Pages. In: ACM SIGMOD International Conference on Management of Data, New York, USA (2003)
5. Cai, D., Yu, S., Wen, J.-R., Ma, W.-Y.: VIPS: a Vision-based Page Segmentation Algorithm. Microsoft Technical Report (2003)
6. Wang, P., You, Y., Xu, B., Zhao, J.: Extracting Academic Information from Conference Web Pages. In: 23rd IEEE International Conference on Tools with Artificial Intelligence, Boca Raton, USA (2011)

7. Zhu, J., Zhang, B., Nie, Z., Wen, J.-R.: Webpage Understanding: an Integrated Approach. In: 13th ACM SIGKDD International Conference on Knowledge Discovery and Data Mining, New York, USA (2007)
8. Nie, Z., Wen, J.-R., Ma, W.-Y.: Webpage Understanding: Beyond Page-level Search. ACM SIGMOD Record 37(4), 48–54 (2009)
9. Yao, L., Tang, J., Li, J.: A Unified Approach to Researcher Profiling. In: IEEE/WIC/ACM International Conference on Web Intelligence, Fremont, USA (2007)
10. Brickley, D., Miller, L.: FOAF Vocabulary Specification, Namespace Document, http://xmlns.com/foaf/0.1/
11. Lafferty, J., McCallum, A., Pereira, F.C.N.: Conditional Random Fields: Probabilistic Models for Segmenting and Labeling Sequence Data. In: 18th International Conference on Machine Learning, Williamstown, USA (2001)
12. Quinlan, J.R.: C4.5: Programs for Machine Learning. Morgan Kaufmann Publishers Inc. (1993)
13. Hand, D.J., Yu, K.: Idiot's Bayes—Not So Stupid After All? International Statistical Review 69(3), 385–398 (2001)
14. Cortes, C., Vapnik, V.: Support-Vector Networks. Machine Learning 20(3), 273–297 (1995)
15. Eddy, S.R.: Hidden Markov Models. Current Opinion in Structural Biology 6(3), 361–365 (1996)
16. McCallum, A., Freitag, D., Pereira, F.: Maximum Entropy Markov Models for Information Extraction and Segmentation. In: 17th International Conference on Machine Learning, Stanford, CA, USA (2000)
17. Forney Jr., G.D.: The Viterbi Algorithm. Proceedings of the IEEE 61(3), 268–278 (1973)
18. Zhu, C., Byrd, R.H., Lu, P., et al.: Algorithm 778: L-BFGS-B: Fortran Subroutines for Large-Scale Bound-Constrained Optimization. ACM Transactions on Mathematical Software 23(4), 550–560 (1997)
19. Wang, P., Xu, B.: Lily: Ontology alignment results for OAEI 2009. In: Proceedings of the Fourth International Workshop on Ontology Matching (OM 2009), Washington, D.C., USA (2009)
20. Wang, P., Xu, B.: Debugging ontology mappings: a static approach. Computing and Informatics 27(1), 21–36 (2008)

# Effective Chinese Relation Extraction
# by Sentence Rolling and Candidate Ranking

Meilun Sheng, Lin Qiu, Chenyang Wu, Haofen Wang, and Yong Yu

Apex Lab, Shanghai Jiao Tong University

**Abstract.** Relation extraction is to discover relations between entities mentioned in the plain text. It can be used to generate semantic data in form of RDF triples representing facts. In this paper, we focus on relation extraction from Chinese text, which is less studied compared with that for English. Chinese words and phrases have great ambiguities on syntax and semantic. Thus, Chinese NLP tools can be insufficient when the sentence is too long or the sentence structure is too complex. Unfortunately, this is the case in the real world data. In order to tackle the limitation of the current Chinese NLP tools, we propose a method called sentence rolling to generate several enhanced inputs from the original input to help generate the correct relation candidates. In order to rank these candidates in an appropriate way, a voting approach is applied based on several statistic-based ranking function. Further, a Relation KB is used to help determine the subject part and the object part for the selected relation candidate. We carried out comprehensive experiments on both real world news corpus and benchmark data combining Chinese Treebank and Chinese Dependency Treebank. The experimental results show that the method can improve the performance of relation extraction significantly compared with the existing ones and cost a reasonable time.

**Keywords:** Relation Extraction, Chinese Relation Extraction, Statistical Method, Dependency Tree, Relation Knowledge Base.

## 1 Introduction

With the information explosion in Internet, Information Extraction(IE) is proposed to overcome the hardness of learning knowledge from the Internet. Relation extraction is a branch of Information Extraction. It is used to discover relations between entities mentioned in the plain text and generate triples describing the semantic information about the relations. It is often involved in building QA system, constructing knowledge base and lots of other applications. For example, DeepQA[9] uses relation extraction as an important component in the system. Therefore, relation extraction is an important task to be solved.

The relation extraction systems are based on varied NLP tools, likes Matetools[2][3]. Chinese words and phrases have great ambiguities on syntax and semantic. Therefore, when the sentence is too long or the structure of the

G. Qi et al. (Eds.): CSWS 2013, CCIS 406, pp. 147–160, 2013.

sentence is too complex, according to our observation, the precision and recall of Chinese relation extraction systems decrease due to the Chinese NLP tools become insufficient. For example, we build a relation extraction system by using Matetools and get an observation that the precision and recall decrease by near 20% when the length of the sentence becomes longer than 35 words comparing with what they are when the length of the sentence is shorter than 25 words. Meanwhile, according to our observation, there are 60% of sentences that are longer than 35 words. It is implied that it is difficult for Chinese relation extraction to work effectively in real world articles.

In this paper, an effective relation extraction method are introduced to overcome the hardness of Chinese relation extraction and improve the precision and recall of existing systems. The proposed method uses existing relation extraction systems as external systems. The proposed method can increase precision and recall obviously when using either of the two different external relation extraction systems and the precision and recall of the proposed method won't be affected obviously by the increasing of the length of sentences. The proposed method introduces the sentence rolling to enhance input for the external systems and generate candidates. It tries to modify the original input to provide sentences which are easy for relation extraction systems to extract correctly. A set of relation candidates is generated by the external system. These candidates are ranked to select the reliable relations. The ranking uses two measurements: One is correctness of candidates. It is measured to ensure the relation quality; The other is semantic completeness. A Chinese knowledge base and a Chinese relation knowledge base are used to measure the semantic completeness of the relations to improve the quality.

The rest of the paper is organized as follows: In section 2, details of some related works are introduced; In section 3, the working flow of the proposed method is introduced; In section 4, two basic relation extraction systems are introduced, and results of experiments about our method and basic relation extraction system are demonstrated. Finally, a conclusion is drown.

## 2    Related Work

Relation extraction is a sub-task of IE to extract relation among relations from unstructured source. The tasks of IE are promoted by MUC and developed by ACE since 2000. Then, studies in IE have shifted to the relation extraction. The well-known relation extraction systems includes Snowball[1], Knowitall[4], Leila[8], Textrunner[13], Statsnowball[14] and Reverb[5]. Technologies such as pattern matching, natural language processing and machine learning are used to solve relation extraction task.

Semantic Role Labelling(SRL) is a special way to extract relations from plain text. It is a multi-way relationship extraction indeed. Arguments such as actor, receiver, location and temporal information will be labelled in the sentence. Matetools[2] provide a multilingual SRL tools which can achieve 0.8476 F1 Score in Chinese. It uses features such as part-of-speech(POS), word sense, dependency relations and so on to identify and classify arguments.

CNFE[12] and other the studies on 5W1H semantic elements extraction[10][11] focus on event extraction from Chinese online news. Event extraction is also a multi-way relation extraction. They use relation classification, pattern matching, SRL and statistical method to extract the relations from news. And they show that mistakes introduced by basic natural language process such as word segmentation, POS tagging and parsing tree can have a strong impact on the result.

Since relation extraction focuses on the relationship between entities, the knowledge of entities can also be helpful to the relation extraction task. Linked Open Data(LOD) can provide a background knowledge of varied entities. The largest Chinese LOD is Zhishi.me[6]. It extracted structural information from three largest Chinese encyclopedia sites and got 5 millions distinct entities.

Based on LOD, a kind of knowledge base focused on the relations is introduced in Junfeng Pan's work[7]. It uses the information provided in Zhishi.me to generate a knowledge base describing varied information about relations such as the domain and range restrictions of specified kinds of relations. This kind of knowledge base can provide a good background knowledge to assist the relation extraction.

# 3   Purposed Method

The relation extraction problem accepts plain text, such as an article, as input and output a group of relations.

Figure 1 shows how the proposed method works. The proposed method requires a relation extraction system as an external component. This relation extraction system is called **the basic relation extraction system**. The basic relation extraction system can be used to extract relations independently. The major purpose of our proposed method is to improve the performance of the basic relation extraction system.

The proposed method can be separated into several sub-processes:

**Data Preprocessing.** This part will do basic NLP on the input, including word segmentation(SEG), pos tagging(POS), name entity recognition(NER), keyword extraction(KE). Then a score called token weight will be calculated. And some key sentences will be selected in the text as the input of relation extraction. This can reduce the total amount of information to be processed. The sentences will be clustered to provide context information for further usage.

**Sentence Rolling.** Sentence rolling algorithm are used to enhance the key sentences. The algorithm provides a lot of enhanced sentences as input of relation extraction. Relation candidates are generated from these enhanced inputs via the basic relation extraction system.

**Candidate Ranking.** Sentence rolling algorithm generates lots of candidates. However, the input enhancement process in sentence rolling can probably introduce some mistaken inputs. Candidate correctness will be calculated to ensure that correct relations extracted from correct input will have higher

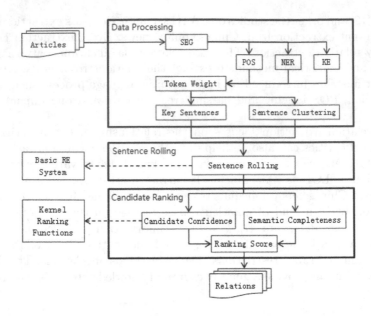

**Fig. 1.** The Working Flow of the Proposed Method

rank. Meanwhile, informative relation candidates will be moved to higher rank by measuring the semantic completeness of the relation. Ranking score will consider candidate correctness and semantic. While calculating candidate correctness, a group of functions called **kernel ranking functions** must be provided according to the basic relation extraction system used.

### 3.1 Data Preprocessing

In Wei Wang's work[10], they suggested a topic sentences identification task to extracted one informative sentence in a news story. In real world articles, an article can have more than one topic. Therefore, their proposed method might be insufficient to multi-topic articles. Meanwhile, the term weight in Wei Wang's work[10] uses tf-idf, which doesn't consider whether the words make contribution to the relation extraction task. Therefore, some improvements are proposed.

In Our proposed method, sentence location, sentence length, number of name entities, title words overlay rate and informativeness of new headline mentioned are the same as those mentioned in the Wei Wang's paper[10]. They are:

$$Score_{loc}(s) = \begin{cases} 1 & i \leq L \\ 1 - log_n i & otherwise \end{cases} \qquad Score_{len}(s) = \begin{cases} 1 & Length(s) \geq C \\ 0 & otherwise \end{cases}$$

$$Score_{ne}(s) = \frac{\sum_{w \in s \cap \{NE_s\}} 1}{Length(s)} \qquad Score_{hs}(s) = \frac{\sum_{w \in H \cap s} TW(w)}{\sum_{w \in H} TW(w)}$$

$$Score_{hne} = \frac{\sum_{w \in H \cap \{NE_s\}} 1}{Length(H)}$$

Token weight is introduced to overcome the weakness of term weight in Wei Wang's work which only uses tf-idf. It is calculated by three stages: In the first stage, the original token weight $TW_o(w)$ is calculated by using these features:

**Name Entity.** The importance of the entity to extract relations depends on the type of entity, i.e. person, location, organization and so on. If the token is a name entity and it's type is $t$, $f_{ne}^t(w) = 1$, otherwise, $f_{ne}^t(w) = 0$.

**Keyword Confidence.** If a word is a keyword, the context around it could be closely related to the main topic of the article. The keyword confidence $f_{kc}(w)$ is normalized into $[0, 1]$

**Verb.** It's hoped that if a sentence is used to extract relations, it has better to own a verb. If the token is a name entity, $f_v = 1$, otherwise, $f_v(w) = 0$.

$$TW_o(w) = \frac{(\sum_t w_{ne}^t f_{ne}^t(e)) + w_{kc} f_{kc}(e) + w_v f_v(e)}{(\sum_t w_{ne}^t) + w_{kc} + w_v} \qquad (1)$$

In stage 2, context score is calculated to estimated whether the context is informative by using the original token weight. A parameter $WindowSize$ should be specified to control how wide the field of view is.

$$ContextScore(w_i) = \sum_{i-\frac{1}{2}WindowSize \leq j \leq i+\frac{1}{2}WindowSize} TW_o(w_j)\frac{|i-j|}{WindowSize}$$

$$(2)$$

In stage 3, the context score is added to the origin token score to smooth the weight score.

$$TW(w) = f(w)\frac{TW_o(w)((\sum_t w_{ne}^t) + w_{kc} + w_v) + w_{cs}ContextScore(w)}{(\sum_t w_{ne}^t) + w_{kc} + w_v + w_{cs}} \qquad (3)$$

Then, the score of the sentence is calculated as:

$$SS(s) = \sum_k w_k Score_k(s) + w_{tw}\frac{\sum_{w \in s} TW(w)}{\max_s(\sum_{w \in s} TW(w))} \qquad (4)$$

To adapt multi-topic articles, not top-1 but top-k sentences are chosen by using score of the sentence. The top-k sentences are called **key sentences**. To prevent the precision decreasing obviously and provide context information when the article is single topic, **related key-sentence set** $RKS$ is introduced.

Let $KS$ be the set of key sentences. For any sentence $S$, $V_w(S) = \sum_{v \in S, v=w} TW(v)$. Maximum spanning tree are generated by using cosine distance space. Then, some edges are cut off. A set of connected components $\{RS_1, ... RS_m\}$ can be gotten.

$$RKS(S) = \{s | s \in KS \wedge s, S \in RS_i, i \in [1, m]\} \qquad (5)$$

$RKS(S)$ is the set of key sentences closely related to the sentence $S$. It can provide context information in the candidate ranking.

## 3.2    Sentence Rolling

According to the observation, if a dependency parser gives a wrong solution for a sentence, it can give a partly correct solution after several remove operations. The solution is partly correct, because it might lose some information of the original sentence. But if the solution can still keep enough information to extract relation which has the same meaning as the original sentence should have, we suggest that the solution should be an **acceptable solution**.

To do the remove operations automatically, a method called the **sentence rolling** is introduced here. The sentence rolling need a sentence, a dependency parser and a basic system doing the relation extraction task as the input. Let $ST(T, N)$ be sub-tree of node $N$. Let $RE(sentence)$ be a function that accepts a sentence and returns a set of relations, this function will use the basic relation extraction. Let $DP(sentence)$ be a function that accepts a sentence and returns the dependency tree of the sentence. For any sentence $sen$, the working flow is listed in Algorithm 1. The algorithm of sentence rolling can easily converted into a multi-thread version to reduce the time cost.

*Algorithm 1: Sentence Rolling*

```
function sentenceRolling(sen) {
    \\Input:
    \\   sen is a sentence with word segmentation, pos tagging,
    \\       name entity recognition information.
    \\Output:
    \\   A set of relation candidates.
    Q = new Queue(); Q.add(sen);
    R = new Set();
    while(!Q.empty()) {
        currSentence = Q.poll();
        R.add(RE(sentence));
        T = DP(sentence);
        n = leftmost sub-node of root node of T;
        rearrange1 = remove ST(n,T) from currSentence;
        Q.add(rearrange1);
        Q.add(remove root node of T from rearrange1);
        for(child : n.children()) {
            rearrange1 = remove ST(child,T) from currSentence;
            Q.add(rearrange1);
            Q.add(remove n from rearrange1);
        }
    }
    return R;
}
```

Sentence rolling will generate lots of candidate events without asking which one is correct or which one is better. It raises the recall as high as possible. The precision will be ensured in the next part. As a kind of byproduct of this part, if a sentence can have sub-relations or paratactic relations, they will be all in the candidate set.

## 3.3  Candidate Ranking

This part introduces two functions used to rank relation candidates. Candidate confidence is to represent the reliability of the given candidate. Semantic completeness is to make the more informative relation candidates have higher rank.

**Candidate Confidence.** There are three kinds of relations among the relation candidates:

1. Acceptable relations.
2. Unacceptable relations generated for wrong input.
3. Unacceptable relations generated for basic system's mistakes.

In the basic relation extraction system, type 2 and 3 are wrong answers. But in our proposed method, these extraction result can also provide information to improve the performance.

**Assumption 1.** A sub-structure $s$ have a high rate of sub-structures appearing frequently in the candidate set, the probability that $s$ is correct is large.

A relation candidate can be separated into one or more parts depending on the basic system used.

**Assumption 2.** The probability that a relation candidate is acceptable is large, if probabilities of all parts of the candidate are large.

Let the origin sentence be $S$. The relation candidate set of $S$ is denoted as $RC(S)$. The part set of a relation candidate $r$ is denoted as $P(r)$. All sub-structure of $p$ are denoted as $SS(p)$. To any sentence $S, r \in RC(S), p \in P(r), s \in SS(p)$, sub-structure $s$ will own a number of tickets $T(s, RC(S))$. The function $T(s, RC(S))$ depends on the basic system used.

For each sub-structure $s$ possibly appearing in $RC(S)$. The popularity of the sub-structure $s$ in the relation candidate set $RC(S)$ is defined as:

$$T'(s, RC(S)) = \sum_{r \in RC(S)} \sum_{p \in P(r)} \sum_{s' \in SS(p)} Sim(s, s') T(s, RC(S)) \qquad (6)$$

$Sim(s, s')$ is a function calculating how similar between $s$ and $s'$. The function $Sim(s, s')$ depends on the basic system used. Let $f(s, p)$ be a punishment function to make sure that the part having large size will not have too many innate advantages. For each part $p$, the confidence of part $p$ is defined as:

$$C(p, RC(S)) = \sum_{s \in SS(p)} f(s, p) \frac{T'(s, RC(S))}{\sum'_s T'(s', RC(S))} \qquad (7)$$

Sometimes, sentence is too short. The number of sub-structures might be not enough to make the ranking reliable. To solve this problem, related key-sentence set calculated in reduction part can provide the context information.

The **candidate confidence** is defined as :

$$Confidence(r, S) = \frac{\sum_{p \in P(r)} C(p, \bigcup_{s \in RKS(S)} RC(s))}{|P(r)|} \qquad (8)$$

$T(s, RC(S))$, $Sim(s, s')$ and $f(s, p)$ are called **kernel ranking function** given as parameters, according to the basic relation extraction system.

**Semantic Completeness.** Because main purpose of relation extraction is to build relations between entities, it is hoped that the subject and object are entities suitable to the domain and range of the relation. Semantic completeness is a measurement of matching degree of the subject and object of the relation.

To simplify the problem and fit our task, all relations just have two types of roles, i.e. subject and object. Assume that subject $a_0$ is independent from object $a_1$ and the type of subject $t_0$ is independent from the type of object $t_1$. $T$ is the set of types of entities. $V$ is the set of types of relations. Semantic Completeness is defined as:

$$\begin{aligned} &Completeness(a_0, a_1, p(r)) \\ &\propto \sum_{t_0, t_1 \in T} \sum_{v \in V} \Pr(a_0|t_0) \Pr(t_0|v) \Pr(a_0|t_0) \Pr(t_0|v) \Pr(v|p(r)) \end{aligned} \qquad (9)$$

Zhishi.me[6] is the Chinese knowledge base which can provide a open classification. The open classification is used as the entity type and $\Pr(a|t)$ can be estimated by counting the entities in $t$. RelationKB[7] is a Chinese relation knowledge base. HIT's synonym dictionary can be used to build set $V$ and estimate $\Pr(v|p(r))$. There is a type UNKNOWN in $T$ for those roles are not entities, $\Pr(a|UNKNOWN) = \Pr(UNKNOWN|v) = 1$, If one of the roles is unknown, the completeness will be punished with a factor.

**Ranking Score.** Candidate confidence is the base of the ranking score. If and only if the relation extracted correctly, the relation can be meaningful. Semantic completeness is used as a punishment factor to move the informative relation to higher rank. The ranking score is defined as :

$$RankingScore(r, S) = Confidence(r, S)(1 + Completeness_{complete}(r)) \qquad (10)$$

## 4   Experiments

### 4.1   Basic Systems Used in Experiments

The proposed method need basic relation extraction system and corresponding kernel ranking functions to work. Two basic relation extraction systems of different styles will be introduced in this section.

**Dependency/SRL Basic System(DSBS).** This basic system is designed as a tree-based system. The dependency tree with semantic role labels is the only part of the relation. Sub-structures are the subtrees of the dependency tree. In this system, Matetools Dependency Parser[3] and Matetools SRL[2] are used to extract relations. A sentence will be sent to dependency parser first and the result will be used as the input of SRL. After SRL, A dependency tree with semantic role label is gotten. Some simple patterns are used to generate relations.

The corresponding kernel ranking functions are defined as:

$$T(s, RC(S)) = \frac{\sum_{w \in s} TW(w)}{||s||} \tag{11}$$

$$Sim(s, s') = \begin{cases} 1 & s=s' \\ 0 & otherwise \end{cases} \tag{12}$$

$$f(s, p) = \frac{1}{||\{n'|d(n', p) \leq d(n(s), p)\}||^{\frac{1}{2}}} \tag{13}$$

**Dependency Path Basic System(DPBS).** This basic system is designed as a slot filling style basic system. The relation will have several parts, a k a slots The sub-structures are the dependency paths.

Dependency path uses the information on dependency tree to generate a shortest path from one node to other node. In this proposed basic system, all paths that start from any nouns, pronouns and verbs, and end with verbs in the sentence will be calculated. To extraction a relation, the basic system classifies all dependency paths in the sentence into subjects, objects and others. The training data uses 1200 sentences chosen in CDTB and CTB and 6500 sentences from real world news.

For any path $s$ in relation candidate $r$, let $T$ be the subtree of start point in the dependency tree of the sentence and $n$ be the end point. $Confidence_{path}(P)$ is the confidence of the path given by the classifier. The ticket of the path is:

$$T(s, RC(S)) = \frac{TW(n) + \sum_{w \in s} TW(w)}{1 + ||s||} Confidence_{path}(P) \tag{14}$$

Path editing distance is used to calculate the similarity function.

- A node should be added/removed. And the added node have the same dependency information as neighbouring nodes. Distance are increased by 0.5.
- The Node is a name entity or a verb, the type of the node need be changed. Distance are increased by 0.6.
- Other kind of changing. Distance are increased by 1.

Let $PED(s, s')$ be the path editing distance between path $s$ and $s'$. And the similarity function is:

$$Sim(s, s') = 0.5^{PED(s,s')} \tag{15}$$

$$f(s, p) = 1 \tag{16}$$

## 4.2  Data Preparation

Two corpus are prepared for experiments:

**Corpus 1.** This corpus is a subset of CTB and CDTB. The restriction of relation is relaxed. A relation can be built even between an entity and a normal
noun or a pronoun. So that, we can have more relations that can be used.
Figure 2 shows the distribution of sentence length sampled in 10000 articles from five Chinese webs. Over 80% sentences are longer than 25 words
and over 60% are longer than 35 words. And the number of verbs in the
sentence can imply the complex level of the sentence. It is necessary to see
the performance of proposed method in different situations. Table 1 shows
the distribution of number of relations and sentences select from CTB and
CDTB in this corpus. $Cxy$ is denoted as the category which is $x$ row and $y$
column in Table 1

**Corpus 2.** This corpus contains 100 articles from web. And each article has
5 informative sentences that are most related to the topics of the article
labelled manually.

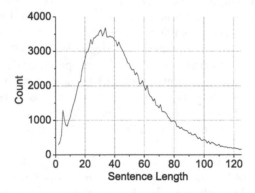

**Fig. 2.** Distribution of Length of Sentences in Real World Articles

**Table 1.** Relations and Sentences Distribution in Corpus 1

| #relation/#sentence | $length < 25$ | $25 \leq length < 35$ | $35 \leq length$ |
|---|---|---|---|
| $\#verb < 6$ | 230/197 | 250/203 | 249/204 |
| $\#verb \geq 6$ | 232/194 | 266/200 | 297/204 |

## 4.3  Evaluation of Key Sentence Extraction

Because relations are only extracted from key sentences. It's important to ensure that the real informative sentences have high sentence scores calculated by
Formula 4. Corpus 2 is used to evaluate the performance of the key sentence
extraction in this section. The proposed method takes top-k sentence as key
sentences. Let $k = 1..8$, and we got Figure 3 that shows the PR curves of the

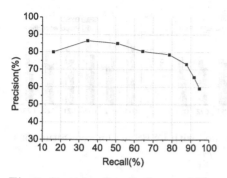

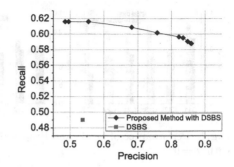

**Fig. 3.** Precision-Recall Curve of Key Sentence Extraction

**Fig. 4.** Precision and Recall of DSBS and Proposed Method with DSBS

key sentence extraction. When $k = 6$, the F-score reaches the maximum 0.6395. When $k > 6$, the precision goes down quickly. Low precision would make the successive part ineffectual. As the conclusion of this experiment, it is suggested that $k$ should be no more that 6, so that most of the extracted key sentences are related to topics of the article and informative.

## 4.4   Evaluation of Relation Extraction

In this section, we evaluate our proposed method on the two basic systems mentioned in section 4.1. Key sentence extraction are disabled, because the corpus have isolated sentences only. Corpus 1 is used to evaluate the method. For each sentence, the ranking score are normalized into $[0, 1]$ by dividing the maximum of ranking score appearing in the candidates. All the candidates have normalized ranking score higher than the threshold $\theta$ are selected as the output relations.

The overall performance of DSBS and our proposed method with DSBS is showed in Figure 4, with $\theta = 0.1, 0.3, 0.5, 0.7, 0.8, 0.9, 0.93, 0.96, 0.99$. The precision is greatly improved when the $theta$ is large, meanwhile, the recall is improved stably. When $\theta = 0.99$, the maximum of F1-score is 0.6984. The recall is very stable. It is 61.61% when $\theta = 0.1$ and it is 58.79% when $\theta = 0.99$. There are two reasons that make the recall so stable:

1. In most cases, there is only one relation in a sentence and the correct answer is the top-1. Therefore, the recall can't be improved obviously.
2. Table 2 shows the error analysis of 100 sentences from which our method doesn't extracted any correct relation. 56% mistakes are generated by mistaken word segmentation or pos tagging. The recall are limited by the mistakes generated in data preprocessing.

**Table 2.** Error Analysis of Proposed Method with DSBS on 100 Sentences

| Mistake Type | Word Segmentation | POS Tagging | Proposed Method |
|:---:|:---:|:---:|:---:|
| Count | 30 | 26 | 44 |

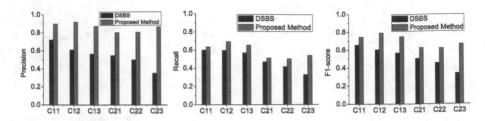

**Fig. 5.** Performance of DSBS and Proposed Method with DSBS in Different Situations with $\theta = 99$

Figure 5 shows the performance of DSBS and our proposed method with DSBS in the 6 situations when $\theta = 99$. Precision and recall are improved in all situations. And they are not affected obviously, like DSBS, by the increasing of the length of sentences.

The situation when comparing DPBS and our proposed method with DPBS are similar. The result are listed in Table 3. Though the performance of DPBS is not very good, we can still gain over 0.2 increment on F1-score by using proposed method with DPBS.

## 4.5 Efficiency Measurement

The proposed method uses sentence rolling to generate lots of enhanced inputs for basic systems to extract relation candidates. This process can cost more

**Table 3.** The Overall Performance of DPBS and Proposed Method with DPBS with $\theta = 0.93$

|          | DPBS   | Proposed Method |
|----------|--------|-----------------|
| Precision | 29.67% | 67.90%         |
| Recall    | 59.65% | 53.54%         |
| F1-score  | 0.3963 | 0.5987         |

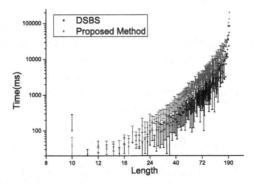

**Fig. 6.** Time cost of DSBS and Multi-thread Sentence Rolling

time than the basic systems can cost. Fortunately, sentence rolling can be easily converted into a multi-thread algorithm. DSBS is used as the basic system in this experiment. Figure 6 shows the time cost when using multi-thread sentence rolling and DSBS. The Multi-thead sentence rolling takes near 3 times time cost than DSBS takes. The multiplying factor basicly depends on the depth of dependency tree.

## 5  Conclusion

In this paper, we propose a relation extraction method using sentence rolling and candidate ranking to greatly improve the performance of basic relation extraction system used as a component in our method with a reasonable time cost. There are four main contributions in our work: First, sentence rolling is proposed to enhance the input for the basic relation extraction system; Second, candidate confidence is introduced into ranking to improve the precision of high rank candidates; Third, semantic completeness is introduced to make high rank candidates more informative; Forth, improved sentence score is introduced to select informative sentences. Several experiments are used to prove the effectiveness of the proposed method. And it is proved that the proposed method can work with varied basic relation extraction systems and improve the performance of them, especially, when the sentence is too long and complex for the basic system to extract relations correctly.

## References

1. Agichtein, E., Gravano, L., Pavel, J., Sokolova, V., Voskoboynik, A.: Snowball: A prototype system for extracting relations from large text collections. ACM SIG-MOD Record 30, 612 (2001)
2. Björkelund, A., Hafdell, L., Nugues, P.: Multilingual semantic role labeling. In: Proceedings of the Thirteenth Conference on Computational Natural Language Learning: Shared Task, pp. 43–48. Association for Computational Linguistics (2009)
3. Bohnet, B.: Very high accuracy and fast dependency parsing is not a contradiction. In: Proceedings of the 23rd International Conference on Computational Linguistics, pp. 89–97. Association for Computational Linguistics (2010)
4. Etzioni, O., Cafarella, M., Downey, D., Kok, S., Popescu, A.-M., Shaked, T., Soderland, S., Weld, D.S., Yates, A.: Web-scale information extraction in knowitall: (preliminary results). In: Proceedings of the 13th International Conference on World Wide Web, pp. 100–110. ACM (2004)
5. Fader, A., Soderland, S., Etzioni, O.: Identifying relations for open information extraction. In: Proceedings of the Conference on Empirical Methods in Natural Language Processing, pp. 1535–1545. Association for Computational Linguistics (2011)
6. Niu, X., Sun, X., Wang, H., Rong, S., Qi, G., Yu, Y.: Zhishi.me - weaving chinese linking open data. In: Aroyo, L., Welty, C., Alani, H., Taylor, J., Bernstein, A., Kagal, L., Noy, N., Blomqvist, E. (eds.) ISWC 2011, Part II. LNCS, vol. 7032, pp. 205–220. Springer, Heidelberg (2011)

7. Pan, J., Wang, H., Yu, Y.: Building large scale relation kb from text. In: International Semantic Web Conference (Posters and Demos) (2012)
8. Suchanek, F.M., Ifrim, G., Weikum, G.: Combining linguistic and statistical analysis to extract relations from web documents. In: Proceedings of the 12th ACM SIGKDD International Conference on Knowledge Discovery and Data Mining, pp. 712–717. ACM (2006)
9. Wang, C., Kalyanpur, A., Fan, J., Boguraev, B.K., Gondek, D.C.: Relation extraction and scoring in deepqa. IBM Journal of Research and Development 56(3-4), 9:1–9:12 (2012)
10. Wang, W.: Chinese news event 5w1h semantic elements extraction for event ontology population. In: Proceedings of the 21st International Conference Companion on World Wide Web, pp. 197–202. ACM (2012)
11. Wang, W., Zhao, D., Wang, D.: Chinese news event 5w1h elements extraction using semantic role labeling. In: 2010 Third International Symposium on Information Processing (ISIP), pp. 484–489. IEEE (2010)
12. Wang, W., Zhao, D., Zou, L., Wang, D., Zheng, W.: Extracting 5W1H event semantic elements from chinese online news. In: Chen, L., Tang, C., Yang, J., Gao, Y. (eds.) WAIM 2010. LNCS, vol. 6184, pp. 644–655. Springer, Heidelberg (2010)
13. Yates, A., Cafarella, M., Banko, M., Etzioni, O., Broadhead, M., Soderland, S.: Textrunner: open information extraction on the web. In: Proceedings of Human Language Technologies: The Annual Conference of the North American Chapter of the Association for Computational Linguistics: Demonstrations, pp. 25–26. Association for Computational Linguistics (2007)
14. Zhu, J., Nie, Z., Liu, X., Zhang, B., Wen, J.-R.: Statsnowball: a statistical approach to extracting entity relationships. In: Proceedings of the 18th International Conference on World Wide Web, pp. 101–110. ACM (2009)

# ZhishiLink: Entity Linking on Zhishi.me

Chenyang Wu, Haofen Wang, Jun Qu, and Yong Yu

Apex Data & Knowledge Management Lab,
Shanghai Jiao Tong University
{wucy,whfcarter,qujun51319,yyu}@apex.sjtu.edu.cn

**Abstract.** Entity linking, which aims to find entities in given text, plays an important role in the trend of shifting from Web of documents to Web of knowledge. In this paper, we present ZhishiLink, an entity linking system targeting the largest Chinese linked open data - zhishi.me. In ZhishiLink, we perform domain-specific disambiguation by leveraging domain topic models to capture the implicit semantics of the entity mentions, in which we collect domains using the categories of zhishi.me. We also evaluate our system on two manually tagged text corpus, namely sina news and sina weibo. Experimental results show that ZhishiLink can successfully resolve most ambiguities raised in both text media with high efficiency. Restful APIs and a web user interface are further provided for external use and user browsing.

**Keywords:** Entity Linking, Zhishi.me, Topic Model, Disambiguation.

## 1   Introduction

There has been a clear trend of moving from a Web of documents to a Web of knowledge in recent years. Wikipedia and its corresponding structured version DBpedia are one of the masterpieces in the world of Web knowledge. Linked Open Data community has also made outstanding contributions to publishing structured data on the Web.

On the other hand, entity linking, as a research topic, has recently attracted much attention from both academia and industry world. It aims to discover entities declared in some knowledge base from arbitrary text inputs. Usually it includes two main steps, namely *entity mention finding* and *entity disambiguation*. Some pieces of text are identified as one entity mention and further linked to some entries in a given knowledge base. With the help of entity linking on the Web of knowledge, it is possible to close the gap between the traditional Web of documents and the vision of Semantic Web.

The main challenges of entity linking on a large knowledge base lies on three aspects. The first one is ambiguity. A mention might refer to different entities in different contexts. For example, "Michael Jordan" can link to more than 20 different entries in Wikipedia. Secondly, an entry in a knowledge base can be expressed in many different forms. This increases the difficulty to identify mentions from different media (e.g., news or microblogging sites) as complete as

G. Qi et al. (Eds.): CSWS 2013, CCIS 406, pp. 161–174, 2013.
© Springer-Verlag Berlin Heidelberg 2013

possible. Lastly, we prefer to link topic-related words correctly and meanwhile completely and discard the common words. Common words are the words which would appear in many scenarios and are not related to the topic of the input document, such as 'place', 'paper'.

In this paper, we present ZhishiLink, an entity linking system on zhishi.me. Zhishi.me is the first effort to publish the Chinese linked open data, and consists of three interlinked encyclopedia sites, including Baidu Baike, Hudong Baike and Chinese Wikipedia. It contains more than 10 Millions of entities from various domains, with complementary information, both structured and unstructured.

In order to tackle the challenges above, the main contributions of ZhishiLink are summarized as follows:

- ZhishiLink utilizes rich structured data in zhishi.me (i.e. redirect links, sameAs assertions, disambiguation pages and anchor texts) as the synonyms of entries, and performs a backward maximal matching algorithm to find possible entity mentions completely
- ZhishiLinks uses domain specific knowledge to identify related topics of the input text and makes linking decisions by considering the most similar candidates. It further filters the topic irrelevant mentions out if no candidate or only candidates of low similarity can be found in the selected domain.
- We evaluate our system on texts from different media, including news text and microblogs, and the result is promising to show the effectiveness of our approach. To extend the usage of our system, we also design a RESTful API and implement a web user interface for ZhishiLink.

This paper is organized as follows. Section 2 introduces related work. We present ZhishiLink in Section 3, including system architecture and disambiguation method. Experiment results on our system are showed in Section 4. Web service API design and user interface are introduced in Section 5. Section 6 concludes our work.

## 2   Related Work

Off-the-shelf entity linking systems stretch across both open-source and commercial world. DBpedia Spotlight[7] is a well-known open-source project towards entity linking on DBpedia, with widely acedemic uses. In Wikipedia-Miner[2], both entity linking and topic extraction are available. Commercial solutions includes OpenCalais[1], AlchemyAPI[2], and these systems focus on robust and full-featured services. OpenCalais endeavors in extracting rich semantic data from text, and provides abundant services with high quality result. AlchemyAPI is a full-stack Natural Language Processing (NLP) service with multifarious services such as sentiment, keywords, and entity extraction, and as a robust platform, is widely used for document analysis. These systems usually target DBpedia or Wikipedia,

---

[1] http://www.opencalais.com
[2] http://www.alchemyapi.com

and mainly dealt with English text. While our work aims to process Chinese and Chinese-English mixed documents.

State-of-art entity linking algorithms usually aim at English, and can be divided into several categories.

**Entity-independent Approaches.** Entity-independent approaches made use of local compatibility based on some context features, and pick the entity with highest similarity. In [8] [2] [3] [11], bag-of-words model and other entity properties like categories, are used to measure the similarity between entity mentions and entity candidates.

**Inter-dependency and Graph-based Approaches.** Unlike local compatibility based approaches, inter-dependency and graph-based approaches took the inter-dependencies into consideration and linking decisions were made based on the other decisions in the document. These methods relied a lot on the inter-dependencies and relationships between the entities, while considering fewer implicit semantic knowledges of the document. Kulkarni et al.[6] noticed the pairwise interdependencies between entity mentions and they converted the problem into an optimization problem, with approximation approaches. Han et al. [5] built a referent graph on all the entity mentions and candidates, then perform an iterative spreading algorithm to obtain stable compatibilities between them.

**Topic and Probabilistic Approaches.** Topic and probabilistic approaches paid attention to the implicit semantics and crytic model underlying the input document. Zhang et al. [10] proposed a method based on topic model and used Wikipedia categories as labels. In [4], Han et al. came up with a probabilistic model, which resolved ambiguity by considering the distribution of entities, possible names of entities and possible context of entities.

From the above, we can conclude that internal dependencies between entities and the implicit semantics of the document are the key elements for entity linking decision. In our method, both of these information are considered to archive a better performance.

# 3   ZhishiLink System

## 3.1   Architecture

The whole architecture of ZhishiLink in shown in Figure 1, with several important enhancements compared to the standard procedure.

The offline procedure involves the following components. *Zhishi.me Service* provides lookup service such as label-to-entity mapping or returning entity structured information such as categories, abstracts, disambiguation pages given the entity's URI. *Domain Extractor* extracts domain instances from entity categories, and passes the instances to *Topic Model Trainer* to train domain topic model. After the models are ready, *Topic Model Service* is used for topic distribution lookup and inference.

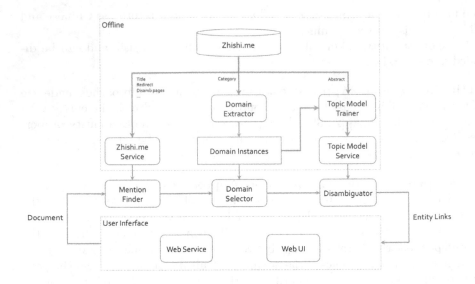

**Fig. 1.** ZhishiLink Architecture

When a document is submitted from the user interface, either by *Web Service* or *Web UI*, it is first processed by *Mention Finder*. *Mention Finder* tries to find entity mentions in the document by looking up the *Zhishi.me Service* for deciding whether a piece of text may refer to some entities. After that, the document and mentions are passed to *Domain Selector* for selecting domains for mentions. *Domain Selector* make domain decisions by checking the domain instances and voting on the mentions' context. The domains are further passed to *Disambiguator* for disambiguation. The *Disambiguator* queries *Topic Model Service* for specific domain models and performs topic distribution and similarity calculation between input documents and entity candidates. Finally, the chosen disambiguated entities are passed to and exhibited on the user interface or returned as Web Service responses.

In the following sections, we will describe each step in detail. First we present how we pre-process our knowledge base triples.

## 3.2   Knowledge Base Structure

Zhishi.me consists of several knowledge bases (KB), including Baidu Baike, Hudong Baike, and Chinese Wikipedia, and they are connected by SameAs connections across different knowledge bases. Of the three KBs in zhishi.me, each KB has its own advantages and disadvantages. For example, Chinese Wikipedia has better quality and more structured information, but relatively small in size, while Baidu Baike has higher coverage. While we perform linking, we integrate all three together, to complement one another.

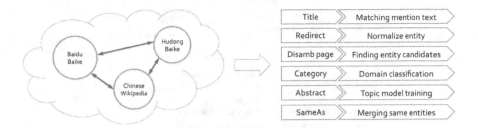

**Fig. 2.** Zhishi.me structure

For each KB of in zhishi.me, collected data and their usage are listed in Figure 2. Titles are used to match the mention text in documents, and redirect links redirect aliases to normalized entities. Ideally, anchor text can also be used for entity mention matching, like in WikiLinks[9]. However, anchor text in Baidu and Hudong are required to be the same as the titles. Disambiguation pages provide more candidates. The domains are extracted from the category information, and models are trained on the abstracts.

Note that our objective entities are the 'unambiguous standard' ones. In a typical KB, disambiguation pages are the navigation pages for entities with the same titles. By 'unambiguous', an entity must not be a disambiguation page. And if an entity has a redirect link, there must be another entity which is the standard form of it and hence it is not 'standard'.

### 3.3  Offline Procedures

The main challenge of entity linking lies on disambiguation. For an entity mention with several linking candidates, we have to choose one of the candidates. We exploit domain knowledge to help the disambiguation procedure.

**Zhishi.me Service.** We further process zhishi.me to provide a service for mention text finding and entity information retrieval. We build a mention-entity mapping as shown in Figure 3. If $a$ is redirected to $b$, $b$ surely has a synonym with title $a$. Also, if $b$ is a disambiguation page, all the entities in page $b$ also have $a$ as their synonym. The above procedure is further enriched by the SameAs connections. Take the example in Figure 3. '交大' will be a synonym of the following entities, 'X:交通大学'(Jiao Tong University), 'X:上海交通大学'(Shanghai Jiao Tong University) and 'X:西安交通大学'(Xi'an Jiao Tong University), in which 'X' stands for any KB source.

We store this mapping and also all the retrieved triples as key-value pairs in BerkeleyDB, and they are used for later lookup.

**Domain Extractor.** Some observation shows that, among all the candidates of an entity mention, these candidates usually do not belong to the same domain. By domain, we are referring to some general fields of knowledge, e.g., Arts,

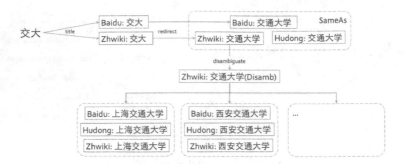

**Fig. 3.** Mention-Entity Mapping

Science, Medical, and so on. Quoting the classical example used in many entity linking works, '*Michael Jordan*' from the sports world are obviously separate individuals from '*Michael Jordan*' from the world of machine learning. Another example would be '*Apple Inc.*' the IT Company and '*Apple*' the food. We make use of this observation and train domain specific models to help our disambiguations. To be specific, we choose 15 different domains, derived on the channels of several famous news site, which are: Sports, Military, Medicine, History, Environments, Entertainment, Education, Culture, Travel, Society, Science, Technology, Arts, Economy, Diet.

While disambiguating, first we perform domain selection for each mention, and choose among the candidates under the selected domains according to their similarities with the input document. Moreover, topic-irrelevant mentions may also be eliminated (linked to NIL) since common words tend to have lower similarity to a concrete domain.

We utilize categories in KB to select domain instances for domain $d$. The categories can be seen as a manual classification result of the KB instances with small granularity. We aggregate on the category throughout all the three KBs and calculate the instance count of each category. Taking the large ones with more than 1000 instances, we get 957 categories. Then we manually classify each of them into domains, with the restriction that each category belongs to at most 2 domains, to restrict common words.

**Topic Model Trainer and Service.** We use LDA (Latent Dirichlet Allocation)[1], a classical topic model to discover the latent topics from a given text corpus, to capture the potential characteristics of each domain. After gathering all the instances for each domain $d_j (1 \leq j \leq 15)$, we train LDA models on their abstracts. Also, we use all the instances from zhishi.me, and train their abstracts to form a background model, which will also be evaluated in Section 4. Generally, abstracts are used since they are comprehensive summaries of the page text, with additional considerations of efficiency.

Two services are provided in the *Topic Model Service* module. One is for looking up the topic distribution of the training instances, the other performs inference on new documents. This component interacts with the online components through HTTP protocol.

## 3.4  Online Processing

In this section, we describe how we process the input document and perform matching and disambiguations.

**Mention Finder.** The first step is segmentation on the input document, resulting a sequence of tokens $t_1, t_2, \ldots, t_n$. Frequently-used matching strategies include maximum forward matching, maximum backward matching and dynamic programming based matching. Experience shows backward matching works better on Chinese text. So we apply the maximum backward matching algorithm, and we treat a token as a word. We scan all the tokens from right to left. For each token $t_i$, find the smallest $j$, such that the text concatenation from $t_j$ to $t_i$ matches some entries in the KB, and consider $t_j..t_i$ as an entity mention, then continue searching from token $t_{j-1}$.

After we confirm all the mention text, the Zhishi.me Service is visited for entity candidates for each mention. Now we have some entity mentions with their corresponding candidates.

**Domain Selector.** We disambiguate in two steps. First we perform domain selection. The purpose is that we can better understand the semantics of the document, and make more reasonable linking decisions under certain domains.

When human read natural language text and encounter an ambiguous entity, we may make use of the context to help our understanding and choose the correct meaning among all the possibilities. For example, 'Apple' with a 'iPad' around has high possibility of talking about 'Apple Company'. On the other hand, 'Apple' with 'Pie' around may be talking about the fruit. This strategy is even better suited for formal text like news, in which only a small portion of the mentions in a context have ambiguity.

Under such observation, Domain Selector uses context information and disambiguation under selected domains can also be seen as disambiguation under context to some extent. We exploit a context based voting algorithm for domain selection by considering all the mentions inside a context window of the current mention. Assume mention $m_i$ starts at token $st_i$ and ends at token $ed_i$, and the context window size is $W$.

$$Context(m_i) = \{m_j \mid [st_j, ed_j] \subset [st_i - W, ed_i + W]\}$$

In our experiment, we take the paragraph the mention is located in as its context. After aggregating the domains of all the candidates in $Context(m_i)$, choose the domains with count larger than the average count. If there is no such domain or none of the candidates belong to any domain, take the background model.

**Disambiguator.** We have known that, for each entity mention, the selected candidate should be related to the selected domains. Under each selected domain, we have two possible strategy. The first one is that we simply keep all the candidates within this domain, and discard all the candidates which are not this domain's instances. The second strategy is to infer the discarded candidates in the first strategy under the domain LDA model, and use the resulting topic distribution in the similarity calculation. We calculate the cosine similarity between each candidate $c$ and the context $t$, using the LDA topic distribution $\theta$ under each selected domain model or the background model, via estimation result or inference.

$$Sim(\theta_c, \theta_t) = \frac{\theta_c \cdot \theta_t}{||\theta_c|| \times ||\theta_t||}$$

For each candidate, pick the maximum similarity under all selected domains. And link the mention to the candidate with the highest similarity, if the highest similarity is greater than threshold $\alpha$. Otherwise link $m_i$ to NIL. We pick $\alpha = 0.1$ in our experiment.

## 4   Experiments and Discussions

In our experiment, we use zhishi.me 3.0 as our target KB, which is constructed from Baidu Baike, Hudong Baike, and Chineses Wikipedia in 2011.

### 4.1   Dataset Preparation

We use manual-annotated data corpus and see how our approaches work on the real web text. Our data corpus consists of two parts, news and microblogging. Each of the two media text has its own properties, in wording, taxeme and so on. Generally, the news text should be more formal, and more full names are used. The microblogging is for fast and short message passing and may use latest words and all kinds of abbreviations, to keep the messages simple and clean.

For news text, we choose 100 pieces of news from Sina News[3]. These news are acquired from several different channels, including Domestic, International, IT, Sports, and Entertainment.

As regards microblogging, we use Sina weibo[4] as our data sources. We perform keyword search on Sina weibo and obtain 200 'weibo's. We use hotspots as keywords, like '以巴冲突' (Palestinian-Israeli conflicts), '苹果' (Apple) and so on. We also consider diversity while choosing weibos, in which we maximize the variety of user backgrounds, including celebrities and normal persons, and so on.

We annotate all these data using a semi-automatic approach. There are three steps:

1. Manual effort are made by finding entity mentions and linked entities. We lookup entities in zhishi.me Lookup Service[5], Baidu Baike and Chinese Wikipedia. If found, we tag this mention with the founded URIs, and consider these URIs as candidates. Note that there may be some entities which

---

[3] http://news.sina.com.cn
[4] http://weibo.com
[5] http://zhishi.me/lookup

are not available in zhishi.me, such as new entries added to Baidu Baike after 2012, we associate those entities with their page URLs.

2. This step tries to find all possible entities in the given text and is necessary because human knowledge is limited and we must utility the help of computer to find all possible candidates. These mentions can overlap with each other, and will be fixed manually in step 3. We use a brute-force entity mention finding algorithm with no disambiguation. After splitting the input text into sentences, we check for every substrings of each sentence. If it matches some entity titles, we put the entity itself as a candidate for this mention. Furthermore, if this entity has a disambiguation page, we use all its disambiguation candidates as mention candidates.

3. Now that, we have two kinds of candidates, manual and automatically generated. One more manual step is needed to check whether each mention is correct and which of his candidates should be linked, involving de-overlapping and manual disambiguation.

By this manual-automatic-manual procedure, we can guarantee to find all the entity mentions. Fairness is further ensured by that, for each piece of text, we have the two manual annotation processed by two different persons, and further checked by a third person.

As mentioned in previous sections, common words is one of the challenges. By common words, we mean words like verbs like '*report*', and common nouns like '*limitation*'. These mentions may link to some entries in the KB, but they mean nothing to people. So we define the mentions we should consider as the *topic-related or user-interested* ones, and use this principle to guide the annotation of the text.

## 4.2   Results

We first evaluate our mention finding algorithm. The result is shown in Figure 4(a). In this experiment, TITLE ONLY uses only entities' titles for mention text matching, without any enrichment of redirects and synonyms. Our Mention Finder uses WITH SYNONYM approach, and successfully find about 10% more mention texts. Mistakes still exist and there are several reasons. Some of the entities are not available in zhishi.me 3.0, and segmentation error would also lead to missing matches. Also, there are some missing expressions in our KB, either with patterns like '美' to '美国'(America), or some abbreviations or aliases, such as 'Big Apple' to 'New York'. We may need to develop a synonym finding algorithm to archive better coverage.

We list all the algorithms used in the disambiguation experiments below. Note that these approaches only differ from disambiguation approaches, while they all use backward maximum matching strategy.

- BLDA is pretty much like the domain based LDA disambiguation, except that BLDA selects the background model every time.

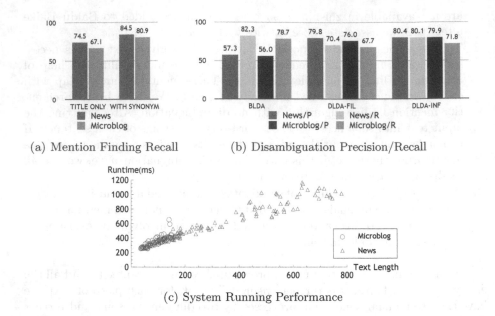

(a) Mention Finding Recall        (b) Disambiguation Precision/Recall

(c) System Running Performance

**Fig. 4.** ZhishiLink Performance and Efficiency Evaluation

- DLDA-FIL uses domain-based LDA disambiguation, and use hard filters which filter out all the candidates outside the current domain. Note that BLDA does not have this filtering process since all the entities are in the background domain.
- DLDA-INF uses domain based LDA disambiguation, and infers topic distribution of the candidates under the selected domain to calculate similarity, instead of the hard-filter strategy in DLDA-FIL.

Table 4(b) shows the result on both news and microblogging text.

In the BLDA approach, the background model is very general to capture the semantic knowledge of the given text. It can perform disambiguation well enough. While the background model is large, it cannot capture the concrete semantic of the document and also links lots of common words. As the result proves our analysis, it has a low precision with a high recall. Particularly, it will find some common noun like '*place*', '*situation*' and fail to disambiguate some entities.

For DLDA-FIL, it shows high precision, which makes sense because we use domain specific model and most of the entities it links are related to some latent topic of the domain model. However, its recall is low due to the domain filter and many mentions are linked to NIL. Because our domain is not be complete, and this will lead to low coverage. In DLDA-INF, The mentions dropped in DLDA-FIL are reused here and use LDA inference to get its topic distribution under the domain model. The result shows high precision and relatively high recall. The recall is not as good as BLDA. Some of the correct entities are dropped. For example, '火箭'(Rockets) in a sports news is certainly talking about '休斯顿火

箭队'(Houston Rockets Team). But our knowledge base does not have 'Rockets' disambiguating to 'Houston Rockets Team' due to incomplete disambiguation page, hence this mention is linked to NIL. This case also evidences that we need more complete synonyms. Compared to BLDA, DLDA-INF correctly linked more than 300 mentions to its correct destination, including linking common words to NIL. Domain-based LDA disambiguation may still link some mentions to wrong destinations. After analyzing the result, we find that LDA may have some redundant topics which mostly contains common knowledge and words, these topics will heighten the similarity of topic-irrelevant mentions and candidates, which is an important problem we will try to solve in our future work.

The result on microblogging data is not as good as news. There are several reasons. As mentioned above, microblogging has text length limitation and many colloquial words are used. For example, in '*Palestinian-Israeli conflicts*', the word 'conflicts' may be omitted, and our synonym-missing problem may get worse under this situation. We may need to explore specific co-reference resolution algorithms for microblogging text. Second, microblogging text provides less context information, and a message can be forwarded by many users with their own comments, resulting multi-topic documents or texts without domain.

Meanwhile, our system is of high efficiency. Figure 4(c) shows a performance statistics. Because our system involves unstable network requests, when communicating with the Zhishi.me Lookup and LDA Service, we run 10 times on each input and take the average run time. We can see from the result, that basically the running time is proportional to the input text length. The average time of each microblogging and news text are 351ms and 699ms respectively.

In summary, our experiments show excellent precision and recall, with high efficiency, while more micromesh efforts are also needed for the methods to work better under different circumstances.

## 5    API and Demonstration

### 5.1    Web Service and API

We have designed a Web Service with RESTful Application Programming Interface (API) for ZhishiLink, allowing programs to access our system. User can use our service via HTTP Post request and the data is in JSON format.

There are several principles throughout our API design.

- RESTful: ZhishiLink severs as a stateless service. No user information is needed.
- Minimal: The API should be simple and fast.
- Representative: The API can provide the exact information which users are interested in, such as mention positions, entity URIs, scores and so on.

Our API accepts JSON input, with a single field 'source', containing the input document text. The output is also in JSON format, which is a list of entity links. Each link is an object containing mention text, linked URIs, start and end position in the document, and finally a score. An example output is shown in Figure 5, for the input string '百度百科和互动百科是中文的两大百科全书'.

```
{
    entities: [
        {
            uris: [ "baidu:百度百科", "zhwiki:百度百科", "hudong:百度百科" ],
            score: 4.494893,
            start: 0,
            end: 4
        },
        ...
    ]
}
```

**Fig. 5.** API Output Example

## 5.2   A Case Study on ZhishiLink

We demonstrate our user interface by an example. Figure 6 shows an overview of our system. We can input Chinese text in the text area in the center, then click Submit.

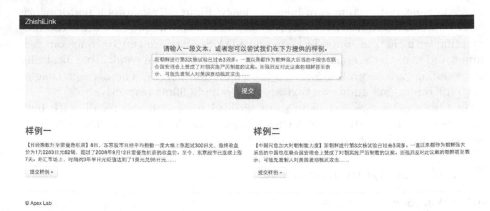

**Fig. 6.** ZhishiLink User Interface

When the text is received and linked by our backend system, the result will be shown as in Figure 7. We can see from the figure that, the linked mention text fragments are surrounded by [], which are also hyperlinks to the corresponding resource page in zhishi.me. The entities are also listed in the left table, with their scores, sorted decreasingly.

We can click an entity in the table, for example, '朝鲜'(South Korea), and all the entities with the same semantic will be highlighted in the right panel. Furthermore, when user hovers on a link on the right, a popup window shows up with the summary text of this entity. Figure 8 shows the abstract of '联合国安理会'(the UN Security Council). We can see from the result, most of the linked mentions are find. There are also some mistakes. The word '严厉'(Stern) is an adjective and should not be linked.

| 实体 | 分数 |
| --- | --- |
| 联合国安理会 | 4.039 |
| 朝鲜 | 1.945 |
| 中国 | 1.922 |
| 美国 | 1.916 |
| 核试验 | 1.200 |
| 核武 | 1.100 |
| 议案 | 1.074 |
| 严厉 | 1.034 |

距[朝鲜]进行第3次[核试验]已过去3周多。一直以来都作为[朝鲜]强大后盾的[中国]也在[联合国安理会]上赞成了对朝实施[严厉]制裁的[议案]。而强烈反对此[议案]的[朝鲜]甚至表示，可能先发制人对[美国]发动[核武]攻击……

**Fig. 7.** ZhishiLink User Interface(submitted)

距[朝鲜]进行第3次[核试验]已过去3周多。一直以来都作为[朝鲜]强大后盾的[中国]也在[联合国安理会]上赞成了对朝实施[严厉]制裁的[议案]。而强烈反对此[议案]的[朝鲜]甚至表示，可能先发制人对[美国]发动[核武]攻击……

**Fig. 8.** ZhishiLink User Interface(abstract)

## 6 Conclusion

Entity linking tries to find mentions in the input text and link these mentions to the corresponding entries in a give knowledge base.

In this paper, we presented ZhishiLink, the first piece of effort for entity linking targeting zhishi.me - the largest Chinese Linked Open Data. We exploited a domain-based LDA disambiguation method, in which topic model are used for domain specific disambiguation and common words elimination.

We paid extra effort to build a manual annotated Chinese text corpus, using a manual-automatic-manual procedure, to ensure accuracy and coverage rate. These data were mostly obtained from news and microblogging. We carried out comprehensive experiments on our corpus, and the results are very promising to show the effectiveness of our approaches. We also provided RESTful API and a user interface to extend our system usage to the fullest.

## References

1. Blei, D.M., Ng, A.Y., Jordan, M.I.: Latent dirichlet allocation. The Journal of Machine Learning Research 3, 993–1022 (2003)
2. Bunescu, R., Pasca, M.: Using encyclopedic knowledge for named entity disambiguation. In: Proceedings of EACL, vol. 6, pp. 9–16 (2006)
3. Cucerzan, S.: Large-scale named entity disambiguation based on wikipedia data. In: Proceedings of EMNLP-CoNLL, vol. 6, pp. 708–716 (2007)

4. Han, X., Sun, L.: A generative entity-mention model for linking entities with knowledge base. In: Proceedings of the 49th Annual Meeting of the Association for Computational Linguistics: Human Language Technologies, vol. 1, pp. 945–954. Association for Computational Linguistics (2011)

5. Han, X., Sun, L., Zhao, J.: Collective entity linking in web text: a graph-based method. In: Proceedings of the 34th International ACM SIGIR Conference on Research and Development in Information Retrieval, pp. 765–774. ACM (2011)

6. Kulkarni, S., Singh, A., Ramakrishnan, G., Chakrabarti, S.: Collective annotation of wikipedia entities in web text. In: Proceedings of the 15th ACM SIGKDD International Conference on Knowledge Discovery and Data Mining, pp. 457–466. ACM (2009)

7. Mendes, P.N., Jakob, M., García-Silva, A., Bizer, C.: Dbpedia spotlight: shedding light on the web of documents. In: Proceedings of the 7th International Conference on Semantic Systems, pp. 1–8. ACM (2011)

8. Mihalcea, R., Csomai, A.: Wikify!: linking documents to encyclopedic knowledge. In: Proceedings of the Sixteenth ACM Conference on Information and Knowledge Management, pp. 233–242. ACM (2007)

9. Singn, S., Subramanya, A., Pereira, F., McCallum, A.: Wikilinks: A large-scale cross-document coreference corpus labeled via links to wikipedia (2012)

10. Zhang, W., Su, J., Tan, C.L.: A wikipedia-lda model for entity linking with batch size changing instance selection. In: Proc. of International Joint Conference for Natural Language Processing, Chiang Mai, Thailand, pp. 8–13 (2011)

11. Zheng, Z., Li, F., Huang, M., Zhu, X.: Learning to link entities with knowledge base. In: Human Language Technologies: The 2010 Annual Conference of the North American Chapter of the Association for Computational Linguistics, pp. 483–491. Association for Computational Linguistics (2010)

# Publishing CLOD of Dangerous Chemicals Based on Semantic MediaWiki

Hailong Deng[1,2], Jinguang Gu[1,2], and Xiaochun Zheng[1,2]

[1] College of Computer Science and Technology,
Wuhan University of Science and Technology, Wuhan 430081, China
[2] State Key Lab of Software Engineering,
Wuhan University, Wuhan 430072, China
simon@wust.edu.cn

**Abstract.** For the problem of integrating massive distributed information about dangerous chemicals and the blank in Chinese semantic knowledge base, this paper proposed a method of constructing CLOD (Chinese Linked Open Data) on that field based on Semantic MediaWiki, making use of the Chinese and English Wikipedia version where each page corresponds to an instance in DBpedia. To make information semantic and machine-readable, we extract instances and metadata from Baidu Baike, add annotations, link to LOD and publish them on the web. So they can be shared and interconnected with other domains. According to experiment, the method proposed in the paper gains benefits of answering rich queries.

**Keywords:** CLOD, Semantic MediaWiki, DBpedia.

## 1 Introduction

With the development of chemical industry, the production and variety of chemicals has grown up at an amazing speed, which has changed and enriched peoples life. There are more than 5 thousand Dangerous Chemicals (DCs), however, among which, nearly 150 to 200 are carcinogens. These DCs share characteristics of combustible, explosive, poisonous, harmful, corrosion, emissive and so on. Casualties, property damage and environmental pollution often happen during their production, transforming, storing and using. Information technology has made a great help to safety management and decision-making of accidents. Encyclopedia, for example, has made a great contribution. Wikis are the most outstanding, like Baidu Baike. This free wiki has been tremendously successful due to the ease of collaboration of its users over the Internet. Its 6 million articles have been written collaboratively by volunteers around the world and almost all of these articles can be edited by anyone with access to the site. However, its search capabilities are limited to full-text search[1], which only allows very limited access to this valuable knowledge base.

At the same time, the shortage of unified representation and formal definition, which results in problem of knowledge reusing, sharing and linking from different systems, has made a large amount of data useless for applications.

G. Qi et al. (Eds.): CSWS 2013, CCIS 406, pp. 175–185, 2013.
© Springer-Verlag Berlin Heidelberg 2013

Recently, focusing on how to integrate structured information and knowledge to answer some semantic queries, mankind have conducted a lot of researches on information integration, at the same time, the industry is also looking for a standard, shared way to build their knowledge base. With the development of semantic web, a large amount of RDF documents have been published on the web. The concept of Linked Open Data(LOD) has attracted huge attention of the academic and engineering. Up to March 2012, the LOD library has 325 datasets, containing over 25 billion triples , covering a number of areas, including medical, entertainment, and geography[2]. The lack of description of Chinese knowledge base, has resulted in that many researchers can not use the data in the LOD directly. Jun Zhao has published part of Chinese Medicine as LOD on the web[3], but its just a small part. Language becomes obstacles to knowledge sharing. LOD construction mainly relies on a large number of areas related ontology technology, however, most of the ontology editor do not support multi-user editing, and not be able to provide comprehensive support for collaboration between domain experts and knowledge engineers. To solve the problem, the paper proposes building CLOD based on Semantic MediaWiki and discusses releasing Chinese DCs knowledge base through LOD.

The work we present in this paper include:

- Collecting information about DCs from Chinese encyclopedias (Baidu Baike) as original data. In total, 1735 instances are parsed.
- Extracting structured information from original data. Metadata should be found to descript information.
- Annotating Pages. Adding Semantic annotations for text.
- To make connections with existing linked data, $< owl : sameAs >$ is used to link resource in CLOD to the same one in DBpedia , a central dataset of LOD.
- A simple search application to show the advantages of semantic information.

## 2   Wiki, MediaWiki and Annotation

A wiki is a website which allows its users to add, modify, or delete its content via a web browser usually using a simplified markup language or a rich-text editor. Most wiki pages are created collaboratively, which results in inconsistency of content because of different background of different people. As the website becomes larger, new problems come out. Like how to find some content immediately or how to share knowledge between different kinds of wikis, etc.

MediaWiki(MW) is a free software open source wiki package written in PHP, originally for use on Wikipedia. The Wiki engine enables each member to search, read, add and edit articles, and thus improve the content of wiki.

Semantic MediaWiki(SMW) is an extension of MW, which aims at enabling reuse of knowledge in Wikis and enhancing search and browse capability. It is a semantic wiki engine that enables users to annotate web content with explicit, machine-readable information. These annotations can be used to show context information to strengthen expression, which provide information to increase navigation ability and semantic search capability .With these semantic information, SMW addresses challenges of traditional wikis:

- **Consistency of content:** The same information often occurs on many pages. How to ensure that information in different parts of the system is consistent, especially as it can be changed in a distributed way?
- **Accessing knowledge:** Large wikis have thousands of pages. Finding and comparing information from different pages is a challenging and time-consuming task.
- **Reusing knowledge:** Many wikis are driven by the wish to make information accessible to many people. But the rigid, text-based content of classical wikis can only be used by reading pages in a browser or similar application[4].

In SMW, each page is divided into a theme group which corresponds to a directory. Each directory represents a set of pages and the first layer directory is the ontology namespace or a knowledge base.In Object-Oriented Programming, class is usually a unified description of multiple objects which have same attribute and behavior,an abstraction of instances. However, in SMW, category is a group of similar pages and instance is a specific class which has some particular attributes and behaviors. A description article page is an instance which will belong to one or more categories. Property is an important metadata which is used to describe the characteristics of instances and can be used to declare the semantic relation between instances. In SMW, most semantic tags correspond to specific ABox statements in OWL DL. As table 1 shows.

**Table 1.** Semantic tags correspond to statements

| OWL | SMW |
|---|---|
| OWL instance | Article page |
| owl:Class | Category |
| owl:ObjectProperty | Relation |
| owl:DatatypeProperty | Attribute |

Annotation or tag is about attaching names, attributes, comments, descriptions, etc. to a document or to a selected part in a text. It provides additional information (metadata) about an existing piece of data, which speeds up searching and helps find relevant and precise information[5]. Semantic Annotation goes one level deeper:

- enriches the unstructured or semi-structured data with a context that is further linked to the structured knowledge of a domain.
- allows results that are not explicitly related on the original search.

In the construction of CLOD of DCs, following steps will be followed:

- **Collect information.** This paper focuses on the domain of DCs and serves the safety management and decision system of accidents. So common sense and management of DCs, handling process and programs of accident should be learned. The common sense of DCs include categories and main characteristics. The management includes production, storage, usage, waste disposal and transforming.
- **Analysis manifestations of DCs knowledge.** After finishing the previous step, domain specific knowledge will be classified and each category is actually a class in OWL. The hierarchical relationships of concept will be defined using parent-child relationship of the classes.
- **Extract core concepts.** Metadata to describe the domain will be found, which is actually the annotation mentioned before.
- **Create instances.** Instances should be created following the classes and annotations defined before.
- **Publish the CLOD.** Publish RDF document on the Web through HTTP protocol following the rules:

  - Content are machine-readable.
  - Significance is strictly defined.
  - Have links to external datasets.
  - Be able to be linked by external datasets[6].

## 3    Information Extraction

We do not extract information from scratch, but from structure existing Web-based encyclopedia information[7]. Our original data source is Baidu Baike. In this section, we will first introduce how to collect original data of DCs and then introduce how to extract metadata.

### 3.1    Collect Original Data

Baidu Baike provides the WYSIWYG(what you see is what you get) HTML editors. So information should be extracted from HTML files. First, download the directory to get name list of DCs. Fig. 1 shows part of names. Second, use network spider to download the HTML pages of each name on that list. Currently, we use python to download 1735 pages including: oxidizers and organic peroxides, compressed gas and liquefied gas, toxic chemicals and infectious substances, corrosive substances, flammable liquids etc.

### 3.2    Extract Metadata

The real use of semantic technologies calls for specialized annotations of complex relations rather than simple and frequent entities such as place, date etc. Users are not willing to look for more than one or two other occurrences of a particular relation that should be automatically tagged. We should analysis establish a set

危险化学品名录[2008版. 国家安全监督管理总局]

| 序号 | 类别 | 项目 | 危险货物编号 | 品名 | 别名 | 英文名 | 英文别名 | CAS号 | UN号 |
|---|---|---|---|---|---|---|---|---|---|
| 1 | 爆炸品 | 具有整体爆炸危险的物质和物品 | 11018 | 迭氮(化)钡〔干的或含水<50%〕 | | Barium azide,dry or wetted with less than 50% water,by mass | | 18810-58-7 | 0224 |
| 2 | 爆炸品 | 具有整体爆炸危险的物质和物品 | 11019 | 迭氮(化)铅〔含水或水加乙醇≥20%〕 | | Lead azide, wetted with not less than 20% water, or mixture of alcohol andwater,by mass | | 13424-46-9 | 0129 |
| 3 | 爆炸品 | 具有整体爆炸危险的物质和物品 | 11020 | 重氮甲烷 | | Diazomethane | | 334-88-3 | |
| 4 | 爆炸品 | 具有整体爆炸危险的物质和物品 | 11021 | 二硝基重氮酚〔含水或水加乙醇≥40%〕 | 重氮二硝基苯酚 | Diazodinitrophenol,wetted with not less than 40% water,or mixture of alcohol and water,by mass | Dinitrodiazophenol | 87-31-0 | 0074 |

**Fig. 1.** Title Part of the names of DCse

of core concepts in the field of ontology conceptual model. Building domain core concepts is building an ontology prototype. DBpedia Extraction Framework[8] is used to extract information from Wikipedia. It embeds all wiki articles in the form of wikitext source and metadata in XML. Here, we take full advantage of infobox template.

Infobox template is intended as a meta-template: a template used for constructing other templates[9]. It is not meant for use directly in an article, but can be used on a one-off basis if required. Fig. 2 shows some information on the page.

The metadata will be used as annotations. These semantic annotations in SMW+ can be well expressed as RDF triple: article is subject, relationship or attribute is property, the target of the link or the value of the attribute is value. Semantic data can be stored in Relational Database, or TDB, ARC as triple. In SWM+, data are stored in database.

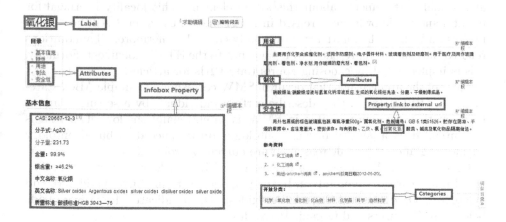

**Fig. 2.** An Example of Resource Page

### 3.3 Import Ontology and Vocabulary

Users are allowed to import or reuse standard and existing vocabulary in Semantic Web documents instead of building each. In order to import one particular vocabulary, the existence of that vocabulary should be confirmed. FOAF,

Dublin Core and SKOS provide some simple and useful vocabularies, which can be checked while building a new vocabulary. At the same time, Falcons, Sindice and some other tools store indice of semantic vocabulary, which can be used to search one particular vocabulary. A common example is importing vocabularies in FOAF. If we want to import foaf:knows, users can add [[imported from:foaf:knows]] to import an external vocabulary[10]. The special property imported from identifies that the attribute come from FOAF vocabulary libs. Also, sameAs, equivalentClass and equivalentProperty in OWL can be used to import equivalent vocabularies. The table 2 shows how they can be mapped syntax in SMW.

**Table 2.** Equivalence mapping in OWL and SMW

| OWL | SMW |
| --- | --- |
| owl:sameAs URI | [[equivalent URI:=URI]](article page) |
| owl:equivalentClass URI | [[equivalent URI:=URI]](category page) |
| owl:equivalentProperty URI | [[equivalent URI:=URI]](relation/attribute page) |

## 4  Annotating Pages

The necessary collection of semantic data in SMW is achieved by allowing users to add annotations to the wiki-text of articles via a special markup. Each article corresponds to exactly one ontological instance, and every annotation in an article makes statements about this single element. This locality is crucial for maintenance: if knowledge is reused in many places, users must still be able to understand where the information originally came. Furthermore, all annotations refer to the concept represented by a page, not to the HTML document. Formally, this is implemented by choosing appropriate URIs for articles.

Most of the annotations that occur in SMW correspond to simple ABox statements in OWL DL, i.e. they describe certain individuals by asserting relations between them, a representable in SMW is intentionally shallow. The wiki is not intended as a general purpose ontology editor, since distributed ontology engineering and large-scale reasoning are currently problematic.

Category in SMW corresponds to class in OWL. Article pages can be classified by adding a category annotation. In the page of describing Silver Oxide, we add [[Category: Oxidizers and Organic Peroxides]] which indicates that Silver Oxide belongs to Oxidizers and Organic Peroxides.

Relation corresponds to Object-Property in OWL. We can describe the relation between two articles by adding tags to the link. In order to express the relationship between Silver Oxide and nitric acid, we can find the link to nitric acid in the article page representing Silver Oxide and add [[Dissolved In: nitric acid]] to construct semantic relation.

Attribute corresponds to DataType-Property in OWL, which will be used to describe the relation between articles and other text. In the page of representing Silver Oxide, we add [[Chemical Formula: Ag2o]] to describe the information of its chemical formula.

Annotations are usually not shown at the place where they are inserted. Category links appear only at the bottom of a page, relations are displayed like normal links, and attributes just show the given value. A factbox at the bottom of each page enables users to view all extracted annotations, but the main text remains undisturbed. It is obvious that the processing of Attributes requires some further information about the Type of annotations. Integer numbers, strings, and dates all require different handling, and one needs to state that an attribute has a certain type. As explained above, every ontological instance is represented as an article, and the same is true for categories, relations, and attributes. This also has the advantage that a user documentation can be written for each element of the vocabulary, which is crucial to enable consistent use of annotations.

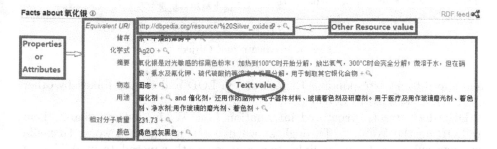

**Fig. 3.** Annotations

# 5   Link to LOD

Most of the time, users edit a certain article based on their personal knowledge which lead to heterogeneous descriptions. Mapping these articles can help to integrate these separated data sources as a whole. We try to break barriers between our Chinese knowledge base and the English one (the LOD).

## 5.1   Data-Level Mapping

Most of the time, users edit some articles by their knowledge. This leads to the description of the inconsistencies between the same concepts. This inconsistency is even more obvious after adding languages. How to map these articles with different descriptions but same content, break language barriers between Chinese knowledge and English knowledge, has become the most important problem. Luckily, the link of $< owl : sameAs >$ has helped to solve that. Equivalence can be marked through add sameAs relations between articles, which result in an connection between articles from cross-platform (Baidu Baike , Wikipedia, Hudong Baike) and cross-language (Chinese, English)[11]. If a Chinese article

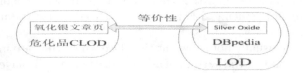

**Fig. 4.** Mapping CLOD

| Predicate | Object |
| --- | --- |
| http://xmlns.com/foaf/0.1/page | http://ontoweb.wust.edu.cn/mediawiki/index.php/氧化银 |
| http://www.w3.org/2002/07/owl#sameAs | http://dbpedia.org/resource/Silver_oxide |
| http://www.w3.org/2000/01/rdf-schema#label | 氧化银 |
| http://ontoweb.wust.edu.cn/mediawiki/index.php/Property:物态 | 固态 |
| http://ontoweb.wust.edu.cn/mediawiki/index.php/Special:Categories | 氧化剂和有机过氧化物 |

**Fig. 5.** Predicate and Object

is mapped to a corresponding English one in LOD and can be linked by other datasets, it is added to LOD cloud.

DBpedia extracts structured information from Wikipedia and shares them as LOD on the Web[12]. Through knowledge extraction framework, DBpedia extracts summary, language links, tags and some other metadata in Wikipedia and stores them as RDF triples. At present, DBpedia is the largest database of cross-domain and one of the most important databases in LOD. All the resources in DBpedia are assigned an URI http://dbpedia.org/resource/Name, where Name is taken from URL of the resource Wikipedia article, which has the form http://en.wikipedia.org/wiki/name. This has created a mapping from a page in Wikipedia to a resource in DBpedia. In creating each instance, we should search the source in Wiki and link through semantic relations. For example, while creating page of Silver Oxide, besides adding semantic marks, we also need to link the page to wikis by adding

[[page::http://baike.baidu.com/view/6875.htm]] and

[[page::http://en.wikipedia.org/wiki/Silver_oxide]],

which will correspondingly link to Baidu Baike, Chinese Wikipedia and English Wikipedia. Also, adding

[[equivalent URI:=http://dbpedia.org/resource/Silver_oxide]]

will link the resource to LOD, which imply that the resource is an instance of LOD. Table 3 show some triples, which the subject is Silver oxide Linking DCs Knowledge Base to LOD is actually a problem of instance mapping. We take full advantage of the internal language links in Wikipedia which describe same theme on different language pages. The source name in DBpedia is the same with one in URL of Wikipedia article pages. So we can follow the following steps to link DCs Knowledge Base to LOD :

- Given a Chinese instance e, find the wiki article e' in Chinese Wikipedia with the same title.
- Find whether there is a cross lingual link between e' and an English article e" in English Wikipedia. If e" exists, get its URL.
- Search the DBpedia URI of e" by looking for the URL.
- Declare URI(e) owl:sameAs URI(e").
- Give unique URI to each resource and publish them on the Web, so that they can be linked by other datasets[13].

### 5.2   Reusing Existing Ontologies

Since SMW is compatible with the OWL DL knowledge model, it is also feasible to use existing ontologies within the wiki. This is possible in two ways: ontology import is a feature that allows users to create and modify pages in the wiki to represent the relationships that are given in some existing OWL DL document; vocabulary reuse allows users to map wiki pages to elements of existing ontologies.

The ontology import feature employs the RAP toolkit for reading RDF document, and extracts statements that can be represented in the wiki. Articles names for imported elements are derived from their labels, or, if no labels are available, from the section identifier of their URI. The main purpose of the import is to bootstrap a skeleton for filling the wiki. Also, ontology import inserts special annotations that generate equivalence statements in the OWL export.

Importing vocabulary allows users to identify elements of the wiki with elements of existing ontologies. For example, the Attribute:name in our inline example is directly exported as the class foaf:name of the Friend-Of-A-Friend vocabulary. Wiki users can decide which pages of the wiki should have an external semantics, but the set of available external elements is explicitly provided by administrator users. By making some vocabulary element known to the wiki, they ensure that vocabulary reuse respects the type constraints of OWL DL. For example external classes such as foaf:name should not be imported as Category.

## 6   Application

The CLOD project focuses on creating semantically enriched structured information, which will help us to answer rich queries[14]. The CLOD of DCs based on Baidu Baike covers production, storage, transport, usage and some other domains, providing a good platform for industrial applications and playing a crucial role in safety management and decision-making of accidents. For example, in order to find out which substance is able to use water to extinguish the fire, we can search the each page of DCs in Baidu Baike to find out the introduction of fire-fighting information. And then map the information with the keyword "water". However, after creating semantic tags for each instance of DCs, we can search the semantic tag "fire-fighting" to map the keyword "water":

{{#ask: [[灭火措施:: *水*]]
| ?灭火措施
| ?化学式
| format=ol | source=wiki
| merge=false |}}

**Fig. 6.** Query

1. 二叔丁基过氧化物 (灭火措施 使用干粉、泡沫、二氧化碳、大量水灭火 化学式 C8H18O2)
2. 六亚甲基四胺 (灭火措施 泡沫、二氧化碳、雾状水、砂土 化学式 C6H12N4)
3. 多聚甲醛 (灭火措施 雾状水、泡沫、二氧化碳、干粉、砂土 化学式 (CH2O)n)
4. 过氧乙酸 (灭火措施 水、泡沫和二氧化碳剂 化学式 C2H4O3)
5. 过氧化二异丙苯 (灭火措施 用大量水覆盖，不要直射，不要接触有机物 化学式 C18H22O2)
6. 过氧化氢叔丁基 (灭火措施 水、泡沫、二氧化碳、砂土 化学式 C4H10O2)
7. 过氧化环己酮 (灭火措施 雾状水、二氧化碳、砂土 化学式 C12H22O5)

**Fig. 7.** Query Results

# 7 Summary

The significance of building CLOD library of DCs based on Semantic MediaWiki is that Chinese DCs knowledge are published on the Web as a standard and open form, breaking the language barriers and playing an important role not only for the entire open knowledge base but also for the domain knowledge building. However, due to the amount and professional restrictions of data source, we should choose more data sources and improve the knowledge with domain experts to make data more reliable. By adding semantic annotations, we can use these information for better solution in production, storage and transport and provide support for decision-making of chemicals accidents.

**Acknowledgement.** This work was partially supported by a grant from the NSF (Natural Science Foundation) of China under grant number 60803160, 61272110 and 61100133, the Key Projects of National Social Science Foundation of China under grant number 11ZD&189, and it was partially supported by a grant from NSF of Hubei Prov. of China under grant number 2009CDA136. It was partially supported by NSF of educational agency of Hubei Prov. under grant number Q20101110, and the State Key Lab of Software Engineering Open Foundation of Wuhan University under grant number SKLSE2012-09-07. It was partially supported by Program for Outstanding Young Science and Technology Innovation Teams in Higher Education Institutions of Hubei Province, China under grant number T201202.It was partially supported by key technologies r&d program of Wuhan under grant number 201110821236.

# References

1. Chernov, S., Iofciu, T., Nejdl, W., et al.: Extracting Semantics Relationships between Wikipedia Categories. SemWiki, 206 (2006)
2. Bauer, F., Kaltenbock, M.: Linked Open Data: The Essentials. Edition mono/monochrom, Vienna (2011)
3. Zhao, J.: Publishing Chinese medicine knowledge as Linked Data on the Web. Chinese Medicine 5(1), 1–12 (2010)
4. Volkel, M., Krotzsch, M., Vrandecic, D., et al.: Semantic wikipedia. In: Proceedings of the 15th International Conference on World Wide Web, pp. 585–594. ACM (2006)
5. Oren, E., Breslin, J.G., Decker, S.: How semantics make better wikis. In: Proceedings of the 15th International Conference on World Wide Web, pp. 1071–1072. ACM (2006)
6. Heath, T., Hausenblas, M., Bizer, C., et al.: How to publish linked data on the web. In: Tutorial in the 7th International Semantic Web Conference, Karlsruhe, Germany (2008)
7. Niu, X., Sun, X., Wang, H., Rong, S., Qi, G., Yu, Y.: Zhishi. me-weaving Chinese linking open data. In: Aroyo, L., Welty, C., Alani, H., Taylor, J., Bernstein, A., Kagal, L., Noy, N., Blomqvist, E. (eds.) ISWC 2011, Part II. LNCS, vol. 7032, pp. 205–220. Springer, Heidelberg (2011)
8. Hellmann, S., Stadler, C., Lehmann, J., Auer, S.: DBpedia live extraction. In: Meersman, R., Dillon, T., Herrero, P. (eds.) OTM 2009, Part II. LNCS, vol. 5871, pp. 1209–1223. Springer, Heidelberg (2009)
9. El-atey, A.A., El-etriby, S., Kishk, A.S.: Semantic Data Extraction from Infobox Wikipedia Template. International Journal of Computer Applications (2012)
10. Becker, C., Bizer, C., Erdmann, M., et al.: Extending SMW+ with a Linked Data Integration Framework. In: ISWC Posters&Demos (2010)
11. Parundekar, R., Knoblock, C.A., Ambite, J.L.: Linking and building ontologies of linked data. In: Patel-Schneider, P.F., Pan, Y., Hitzler, P., Mika, P., Zhang, L., Pan, J.Z., Horrocks, I., Glimm, B. (eds.) ISWC 2010, Part I. LNCS, vol. 6496, pp. 598–614. Springer, Heidelberg (2010)
12. Auer, S., Bizer, C., Kobilarov, G., Lehmann, J., Cyganiak, R., Ives, Z.G.: DBpedia: A nucleus for a web of open data. In: Aberer, K., et al. (eds.) ISWC/ASWC 2007. LNCS, vol. 4825, pp. 722–735. Springer, Heidelberg (2007)
13. Wang, Z., Wang, Z., Li, J., et al.: Knowledge extraction from Chinese wiki encyclopedias. Journal of Zhejiang University Science C 13(4), 268–280 (2012)
14. Herzig, D.M., Ell, B.: Semantic mediawiki in operation: experiences with building a semantic portal. In: Patel-Schneider, P.F., Pan, Y., Hitzler, P., Mika, P., Zhang, L., Pan, J.Z., Horrocks, I., Glimm, B. (eds.) ISWC 2010, Part II. LNCS, vol. 6497, pp. 114–128. Springer, Heidelberg (2010)

# Toward Ontology Representation and Reasoning for News

Xubo Wen[1,2], Xiaoli Ma[2], Juanzi Li[2], Jeff Z. Pan[3], and Jiayu Xie[1]

[1] Information Engineering Institute of Technology, Naval Academy of Armament, P.R. China
[2] Department of Computer Science and Technology, Tsinghua University, Beijing, P.R. China
[3] Department of Computer Science, University of Aberdeen, UK
{wenxubo,thumxl,lijuanzi2008}@gmail.com,
Jeff.z.pan@abdn.ac.uk, wenbruce@163.com

**Abstract.** Most research work on news mining nowadays covers phrase and topic level. A few works conducted on logical level mainly focus on personalized news service and no special efforts are put on the applications of ontology techniques on deep news mining. In this paper, we demonstrate a whole strategy for deeply understanding event-focused news taking the advantage of ontology representation and ontology reasoning. We propose an ontology-enriched news deep understanding framework ONDU which addresses the following problems: (1) how to transfer parsed news content into logical triples by using domain ontology. (2) The application of ONDU based on the reasoning results from the ontology reasoner TrOWL over the RDF data expressing the news. Through this whole strategy we can detect the inconsistence among multiple news articles and compare the different information implied in different news. We can even integrate a set of news content through merging the RDF data. The empirical experiment conducted on news from several portals shows the effectiveness and usefulness of our method.

**Keywords:** Ontology, reasoning, news mining, text understanding, TrOWL.

## 1   Introduction

The rapid development of the World Wide Web prospers the news-related data diffusion and spread over the Internet at great pace. The research on news mining has gained a lot of attention from both industries and research communities due to the prevalent demands of web users accessing to various information around the world over proper span of time. Generally speaking, there are two kinds of research around news mining: topic level exploring over single source of news articles or news from different sources; phrase level analysis for finding some information among news such as contents diffusing patterns on the Internet. The former kind can be divided into several sub-categories: the presenting of the synthesized multiple dimensional information including news, named entities, background information linking from knowledge base, comments, microblogs, and so on, which is often organized based on topics or events [1]; topic detection and tracking problems is specified to acquire the

G. Qi et al. (Eds.): CSWS 2013, CCIS 406, pp. 186–198, 2013.
© Springer-Verlag Berlin Heidelberg 2013

latent topical structure of similar news tracking over time and across languages [2]. The second kind of work is focus on discovering the theme patterns or tracking short, distinctive phrases or retrieving and ranking sentences related to the prediction of future events by text mining methods.

Ontologies provide a structured, semantically rich way of modeling a domain. The ability to reason over relationships defined in ontology, therefore the related instances to their abstracted types is the primary benefit to using ontologies [3]. The domain model was represented as ontology, providing semantic meaning to the domain model. Most ontologies represent content or conceptual domains. Ontologies provide the semantic relationships between objects and instances in a particular domain. As we know, one news report can express one or several aspects of an event and different news reports may cover different aspects of the same event. For example that every company has an annual financial report in the same schema each year, people may only care about the changes which comprise a little fraction of the report. The surface meaning of news articles is easy to speculate, but detecting the false and implied information needs further comparison and comprehensive reasoning which consumes a lot of human labors. The goal of this paper is to semantically integrate the news of different media to get a comprehensive understanding in a machine way.

There are two main challenges to Represent the news articles ontologically, analyze the semantic form triples logically.

Our contribution includes two parts. Firstly, we propose a systematic framework for textual deep understanding which can perform integration and reasoning on news. Secondly, we provide a method to automatically generate the ontological representation from news articles and their application in news comparison, conflict detection and combination. There are many practical advantages of this proposal, for example that the theory can be used in XX Equipment Management Integrated Information System providing some kinds of intelligent managements. Our application goal is detecting the conflicts, differences and integrating different news on the same theme to get comprehensive news service for users while the usual similarity related strategies cannot provide this form of information fusion.

The rest of the paper is organized as follows. Section 2 explains the related work. In section3, we describe the model including the framework and the functions. In Section 4 and section 5 we expose the experimental results and evaluation. Finally Section 6 concludes the paper.

# 2    Related Work

There are a number of tools to help people better understanding news articles around the world and most of them aim at using algorithms to reorganize and filter news [4]. Historical news reports of the same kind of events can provide people a lot of new perspectives of looking into future as future-related topics can make out domain-specific predictions. It is a crucial way to understand how events will develop for a given topic, thus some work has been done on ranking related news over time [5]. Different to traditional news recommendation systems which consider both of user and news content information, nowadays news recommendation combines new features such as access patterns, named entities, popularity and timeliness of news articles [1]. Topic detection and tracking (TDT) aims to devise powerful, broadly useful, fully automatic algorithms for determining the topical structure of news streams [2]. Such automatic discovery and threading of topic could be quite valuable in many applications where people need timely and efficient access to large quantities of in-formation. Grouping similar news articles and linking the clusters over time into events is a common way of topic detection and tracking. Long time view on how given story or event developed over time besides the latest news can supply users with synthesized accurate description of events [6]. EventSearch demonstrated a system for event extraction and retrieval on four types of data: web news articles, newspapers, TV news program, and mirco-blog short messages which show people the way of integrating information from multiple data sources [7]. NewsMiner is a system centered on events concentrating on reconnecting news and background information over linking messages based on human intuition of news browsing behaviors [8].

The concept and related techniques of ontology are for Internet evolving into semantic web which characters with deep understanding of information. Before the machine can understand any raw text, we have to segment and preprocess the text into semantic component for deep understanding and analysis. Through integrating ontology schema and structured text together, we can get the extended understanding of the news articles and background information. There are several efforts making news services taking advantages of ontology techniques. [9] They use ontology-based reasoning to build user profile in the domain of research. Our goal is to integrate multi-source news and detect conflicts between the news of different sources through ontology reasoning engine.

In [10], they present the semantic structure of a document as a semantic graph for further document summarizing. It adopts the deep syntactic analysis results, that is, subject-predicate-object form as the nodes and edges of semantic graph. Linguistic tool NLPWin is applied to achieve several tasks: extracting logical form triples, resolving pronominal reference, identifying the named entities through surface form matching and text layout analysis. Besides, WordNet is used to normalize the terms and merge the semantic graph for further analysis.

The ontology is categorized into two forms, a heavyweight ontology providing meanings for things and relationships, a lightweight ontology with little semantic meaning between the relationships. Our ontology provides a domain model rich in semantic meaning and relationships. For the experiment we manually created a heavyweight ontology using Protégé, then populated it with instances of the classes and properties extracted from news logical triples, and exported into the OWL files. Building the Experimental Ontologies ----Since we need to reason over the relationships between concepts and instances, there is only one choice for building ontology—heavyweight ontology and we build it manually.

We use the ontology reasoner to reason over the different sources news content represented as instances and properties of well-designed ontology, for checking the inconsistence and comparing the contents of different source news.

# 3    Methodology

In order to acquire the deep semantic understanding of news, we firstly extract the logical form triples from news text starting with syntactic analysis. They based on the structure of the ontology; we populate the domain ontology with the logical form triples for further reasoning. By analyzing the relationships between instances and classes, we identify the same facts and the different facts between different news, and then compute the conflicts or merge them into more comprehensive descriptions of the news.

We propose a systematic framework shown in Figure 1, aiming at capturing deep meaning of news reports through combining and extending the known methods. There are mainly four phases, each part extended in details in the following subsections:

- Preprocessing – We adopt MSTParser to parse text and extract logical form triples, including resolution for named entities and pronomial anaphora resolution.
- Ontological data extraction – We build the ontology schema based on the domain information extracted from Wikipedia, cyc, wordnet. Later the we populate the ontology with logical form triples.
- Reasoning – We use ontology reasoning engine TrOWL[1] to reason over the completed ontology.
- Analyzing – Based on the ontology and reasoning outcomes we can check the conflicts of different news, compare the differences among multiple news, combine different news content for further use.

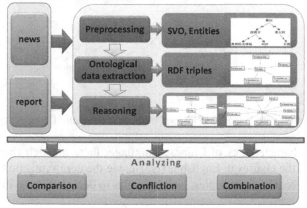

**Fig. 1.** Framework of the model

## 3.1    Preprocessing

We apply deep syntactic analysis to each sentence of news for extracting logical form triples, subject-predicate-object, using MSTParser[2] trained on Semantic Dependency

---

[1]    http:// trowl.eu /
[2]    http://www.seas.upenn.edu/~strctlrn/MSTParser/MSTParser.html

Network [9] which is a large Chinese semantic dependency corpus. We apply pronoun resolution, entity recognition and normalization to refine the set of triples.

- Named Entity Recognition and Normalization– Named entities give the details of 5W1H, which consists of people, locations, organizations and time. We identify named entities through the approaches proposed in [10] by linking them with the knowledge bases, aiming at consolidating expressions that refer to the same named entity and forming a basis classes and hierarchical relationships for ontology schema building.
- Pronominal Reference Resolution – We use MSTParser syntactic and semantic labels to trace and resolve pronominal references as they appear in the text. Pronoun reference resolution within a single sentence is performed in NLP MSTParser automatically. As for pronominal references that cross the sentence boundaries, we apply the approaches proposed in [10].
- Deep Syntactic Analysis – After the above two process, we begin deep syntactic analysis of the text. We first segment the news text into individual sentences, convert sentence text into a dependency parse tree (Figure 2) that represents the syntactic structure of the text using MSTParser. Then we extract the core syntactic parts to form a sentence logical form that reflects the main meaning, semantic structure of the text. Based on the dependency relationships confined in part complex syntactic units, we transform part complex syntactic units into additional logical form triples according to several rules below.

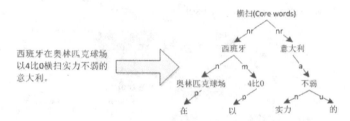

**Fig. 2.** Sample results from MSTParser trained on Semantic Dependency Network

Due to the characteristics of Chinese language, there are rich rhetorical relationships instead of the clauses in English texts. Beside the core syntactic components can be processed as logical form triples, we need rules to deal with the remnants.

- Rule 1:   Appositives can be transformed into class and individual relationship which can be added into relevant ontology schema part as is-a relation directly. For example, in "卫冕冠军西班牙4比0完胜意大利" (The Spain, the defending champion, outperformed the Italy completely at the score of 4: 0.), "卫冕冠军" (the defending champion) can be defined as a class   and "西班牙" (the Spain) as an instance of it.
- Rule 2:   Complex objects in the form of (s1, v1, o1 (s2, v2, o2)) can be transformed into (s1, v1, o1) and (s2, v2, o2). For example, in "西班牙追平德国3次夺冠的纪录" (The Spain tied the record which the German has get champion 3

times.), we can transform the sentence into "西班牙追平德国" (The Spain tied the record) and "德国3次夺冠" (The German has get champion three times.).

— Rule 3:   Some stop words are to be discarded directly, such as auxiliary, interjection and so on..

## 3.2   Ontological Data Extraction

The target of this part is to parse triples to RDF triples, which consists of two components. First we build the ontology schema with domain information from several sources. Then we populate the domain ontology with logical form triples. In fact, the two parts often iterate since we have to consolidate the named entities within ontology schema and news texts.

To guarantee the accuracy and authority of the ontology schema, we merge the information of several knowledge bases, such as Wikipedia, CYC, and WordNet, on given domains. Our ontology schema mainly bases on Wikipedia and the other knowledge bases are used to complement the information. For the experiments, we choose football games and financial reports.

In order to apply to all the events in the domain, the ontology should cover almost all the aspects of the domain. We build the unified schema for the chosen domain either from domain experts or from now existing knowledgebase. The schema includes class and object property, the class is composed of the entities, and the object property is constructed by the predicate. We extract the hierarchical relationships as the infrastructure of our ontology schema. Simultaneously we cover all the instances that appear in the news. As for the properties, we project several verbs of the same meaning into one standard one as their parent property node according to synonym dictionary. We also cover all the predicates in news texts. At last, we check the ontology manually.

For experimental purpose, we build the schema in the domain of football match, which contains the time when it happens, the place where it is held, the players who score or assist, the time when it scored, the coach, the team members and son. The schema covers all the contents in the domain.

We populate the ontology with our refined logical form triples. Actually after preprocessing, we basically get the main classes and instances and properties. The next step is to relate the properties with class information, especially the individual instances.   We process the triples one by one (Figure 3).

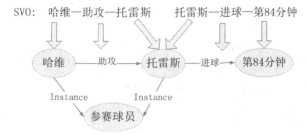

**Fig. 3.** Parsing triples to RDF triples

Based on the schema, we transfer the formalized triples to the ontological data semi-automatically with limited human intervention. Two steps are needed: mapping the entities to the classed defined in the domain ontology, mapping the triples to object properties. The dataset of the entities will be mapped to the members of the class in ontology. But how to map the entity to the proper class is a challenging problem. That's why we adopt the most extensive knowledge of dataset Wikipedia; there is almost all the explanation for every word. The class description of the ontology can be gotten from the Wikipedia. For example, the interpretation of a football team includes all the members who have different roles and positions in match, the coach, the history etc. All the information is related to a football match. Though the football match, like any other matches, is full of complex concepts, e.g., off-side, penalty kicks, different kinds of fouls, etc., we only reasoning over the aspects that most people are interested in and appear in sports news reports.

The triple structure of SVO (subject–verb–object) corresponds to different part in schema. The class of the subject is mapped to the domains in ontology, and the class of the object is mapped to the ranges in ontology. The classification is decided by the Wikipedia. Because of the multiple expressions of a word in Chinese, the object properties need to contain almost all the expressions in the domain. For example, "team A beat team B", in which "team A" and "team B" are the member of the class team, "beat" is in object property, so the relationship is (team, beat, team) .

### 3.3    Reasoning

The ability to reason over relationships defined in an ontology and related instances to their abstracted types is the primary benefit to using ontologies [3]. Our goal is to compute the differences and inconsistence by machine automatically.

Much shallow information from the news is easy to get, but the implied information need to be reasoned. For instance, people will die and Socrates is a man, we can reason Socrates will die.

It is impossible to compute the relationship by using the raw texts only, but we can reason the extra relationships over the triples confined to the well-built ontology and compute the relationships among different news. Especially we can compare the similarities between different texts by many state-of-art models, but we cannot detect even the minor conflicts and differences among the texts. The ontology reasoner TrOWL offers tractable support for all the expressive power of OWL by using quality guaranteed approximate reasoning, which can help us compute the relationships.

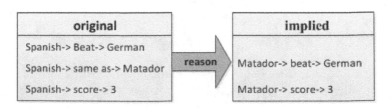

**Fig. 4.** Reasoning results

With the reasoner, we can get three types of the reasoning results from the ontology, which are atomic concept subsumption, atomic class assertion and atomic object property assertion. An example is show in figure 4. From news reports, we get the original logical form triples on the left and implied logical form triples on the right. The reasoning procedure takes all properties of Spanish and applying to Matador since they are appositives. The appositives are formed into "same as" relationship.

### 3.4    Analyzing

There are three cases for complete understanding of the news contents. Firstly, we get the differences from the reasoning results directly over the ontology. Secondly, we combine these differences into an integrated news OWL for queries and browsing, because fusion of multi-source news content can help us understand the news event in a more comprehensive way, while one specific news report often focuses on one or several aspects of the event and different news reports may provide different perspectives for scrutinize into the event. Thirdly, the conflicts will be detected to guarantee that we can get the right image of the reported event.

The comparisons of different news consist of three aspects: are the instance individuals that appear in different news same? Are the predicates same? How many different reasoning results? We do this on the basis of the reasoning results as owl reasoner can provide many enriching semantic relationships basing on knowledge bases behind.

## 4    Experiment

We use the news on the 2012 final UEFA European Football Championship match as a case study and show our empirical experiment result below.

The example news reports are chosen from Chinese News Web and Sina Sport News.

Firstly, we convert the content of the news to triples and construct ontology using protégé 4.1.0. Part of the results is showed in Table 1, where News A and B represent two different reports. For better understanding our analysis results, we describe our ontological triple in English

**Table 1.** Converting the news to triples (part)

| News | Subject | predicative | Object |
|------|---------|-------------|--------|
| News A[3] | Harvey | Goal-Time | The 14th minutes |
| | Harvey | Kick-Position | Central field |
| | Harvey | Kick-Way | Pass ball |
| | Harvey | Kick-way | Straight |
| | Iniesta | Kick-Direction | Right side of the restricted zone |

---

[3] http:// www.qzwb.com/gb/content/2012-07-02/content_4024006.htm

**Table 1.** (*Continued*)

| | | | |
|---|---|---|---|
| | Iniesta | Pass | Giorgio Chiellini |
| | Silva | Same-as | D-Silva |
| | Silva | Kick-Position | 6 meters in front of the goal line |
| | Silva | Kick-Method | Head |
| | Silva | Score | In goal door |
| News B[4] | 14 minutes | Goal-Time | Type |
| | Cesc Fabregas | Assists | Silva |
| | Silva | Same-as | T-Silva |
| | Silva | Kick-Method | Head |
| | Silva | Score | Burst open the goal door |

Secondly, we get more relationships from triples by using TrOWL reasoning over the completed ontology, which are showed in Table 2.

**Table 2.** Reasoned new triples

| News | Subject | Verb | Object |
|---|---|---|---|
| News A | D-Silva | Kick-Position | 6 meters in front of the goal line |
| | D-Silva | Kick-Method | Head |
| | D-Silva | Score | In goal door |
| News B | T-Silva | Kick-Method | Head |
| | T-Silva | Score | Burst goal door |
| NewsA + NewsB | D-Silva | Same-as | T-Silva |
| | In goal door | Same-as | Burst open the goal door |

Thirdly, we compare the news through the original and reasoning triples from the NewsA and NewsB, as showed in Table 3.

**Table 3.** The Comparison of the results

| News | Subject | Verb | Object |
|---|---|---|---|
| NewsA – NewsB* | Harvey | Kick-Position | Central field |
| | Harvey | Kick-Method | Pass ball |
| | Iniesta | Pass | Giorgio Chiellini |
| | Silva | Kick-Position | 6 meters in front of the goal line |
| NewsB – NewsA* | Cesc Fabregas | Assists | Silva |

---

[4] http://www.chinanews.com/ty/2012/07-02/3999358.shtml

NewsA – NewsB means the triples which only exsit in NewsA but not in NewsB, vice versa.

Finally, based on the differences, we can get the comprehensive triples which are showed in Table 4.

**Table 4.** The comprehensive result

| Subject | Verb | Object |
|---|---|---|
| 14 minutes | Goal-Time | Type |
| Harvey | Kick-Position | Central field |
| Harvey | Kick-Method | Pass ball |
| Harvey | Kick-Method | Straight |
| Iniesta | Kick-Direction | Right side of the restricted zone |
| Iniesta | Pass | Giorgio Chiellini |
| Cesc Fabregas | Assists | Silva |
| Silva | Kick-Position | 6 meters in front of the goal line |
| Silva | Kick-Method | Head |
| Silva | Score | In goal door |
| D-Silva | Same-as | T-Silva |
| Silva | Score | In goal door |
| In goal door | Same-as | Burst open the goal door |

After scrutinizing the comparison results, we discover that news B puts more attention on the details of match itself while news A inclines to criticize the match process.

**Table 5.** Statistics of types and relationships

| r | A | B | A-B | B-A | A^B | A+B |
|---|---|---|---|---|---|---|
| types | 70 | 47 | 26 | 3 | 44 | 73 |
| relationships | 36(21) | 42(34) | 32 | 37 | 6 | 75 |

(A, B are two pieces of news; A-B: terms appear only in A; B-A: terms appear only in B; A^B: terms appear in both pieces of news; A+B: the total number of terms in merged form.)

The comparison results of types and relationships after reasoning are showed in table 5. The results demonstrate the effectiveness of our method which can detect the conflicts, compare and merge news from different sources focusing on the same event. The experiment also shows that the method can substitute part of labor work.

# 5   Evaluation

The state-of-art evaluation approaches on ontology techniques are listed below [11]: (1) those based on comparing the standard ontology which can be considered as a template; (2) those based on evaluating the application results of the ontology; (3) those based on comparing the ontology with the source data covered by the ontology; (4) assess how well the ontology meets the predefined criteria, standards, requirements, etc. Most evaluations are performed in two stages, an internal evaluation during the constructing process, and an external evaluation by the clients or experts after the construction is completed [12].

In our experiment, we mainly perform evaluations on the application targets including the reasoning results and application results. Due to the reasoner TrOWL, we can check the inconsistence within the ontology and obtain new logical relationships. As our experiment data does not contain inconsistence cases, we add several artificial conflict logical triples into the original data to check the efficient of our proposed framework.   The result is showed below.

After the combination and comparison of different news contents, we apply human judgments into evaluation of the analysis results. To evaluate the differences and the comprehensive results from two news with the same topic both of which contain at least 1000 Chinese words, we get at least three volunteers who familiar with the domain to evaluate the results as there are no tools to accomplish this kind of task and no benchmark to reference. The differences are clear at a glance, especially if the data including number, date, name and so on, while the differences on semantics can't be identified directly. The evaluation result also proves that the merged report can help people understand the news comprehensively. Most importantly, all of the processes can accomplish in a short time.

**Table 6.** Evaluation results

| Terms | Original relationships | | Reasoned relationships | | | | |
|---|---|---|---|---|---|---|---|
| data | News A | News B | News A | News B | C | D | M |
| | **27** | **39** | **41** | **47** | **7** | **74** | **81** |
| User A: % | 77. 8 | 79. 5 | 78. 0 | 78. 7 | 85. 7 | 77. 0 | 77. 8 |
| User B: % | 85. 2 | 84. 6 | 85. 4 | 83. 0 | 85. 7 | 83. 8 | 82. 7 |
| User C: % | 81. 5 | 82. 1 | 80. 5 | 80. 9 | 71. 4 | 82. 4 | 81. 5 |
| average: % | 81. 5 | 82. 1 | 81. 3 | 80. 9 | 80. 9 | 81. 1 | 80. 7 |

(Common terms—c, differences—d,: respectively the numbers of common and different terms   between two pieces of news; merged—m: the number of terms in the merged form of news. )

Through comparison, we can find that two pieces of news are completely consistent on the basic aspects, such as match place, time, match results, people, and so on, while one mainly focuses the details, such as the scoring time, angles, distance, and so on, and the other one mainly focuses on critics of the match, such as the honor of the players, the performances, and so on.

Our evaluation results are showed in table 6. We organize the logical form triples into a questionnaire and ask the volunteers to select which triples are correct and informative based on the raw news. After that, we integrate the data into table. From the table we can formulate a view that our framework can replace human labor up to a great amount. Through analyzing the results, we found that the differences between human and our method are mainly due to the particular characteristic of Chinese grammar.

## 6    Conclusion

We transform the original news into a set of triples. By building the schema at first, we then construct the ontology through populating it with formalized triples.  Using ontology reasoner TrOWL we reason over the ontology for more deep information. Comparing the news focusing on the same event, we can find the differences and detect conflictions. The combination of different OWL files derived from different news confined to the same ontology schema, we can get the comprehensive information around the same theme.

In this paper, we propose a systematic framework can deeply understand the text, and we implement a method to generate the ontological representation automatically. At last, we perform news comparison, conflict detection and integration semantically. The experiment demonstrates that the strategy gets the results automatically and quickly.

In the future, we will improve the performance of news preprocessing and Learn more transformation rules automatically or semi-automatically. Because we only apply the framework in football match, we want to perform in more general domain.

**Acknowledgement.** The work is supported by the Natural Science Foundation of China (No. 61035004, No. 60973102), 863 High Technology Program (2011AA01A207), European Union 7th framework project FP7-288342, and THU-NUS NExT Co-Lab.

## References

1. Li, L., Wang, D., Li, T., Knox, D., Padmanabhan, B.: Scene: a scalable two-stage personalized news recommendation system. In: SIGIR 2011, pp. 125–134 (2011)
2. Wayne, C.L.: Multiligual topic detection and tracking: successful research enabled by corpora and evaluation. In: Conference: Language Resources and Evaluation - LREC (2000)
3. Conlan, O., O'Keeffe, I., Tallon, S.: Combining Adaptive Hypermedia Techniques and Ontology Reasoning to Produce Dynamic Personalized News Services. In: Wade, V.P., Ashman, H., Smyth, B. (eds.) AH 2006. LNCS, vol. 4018, pp. 81–90. Springer, Heidelberg (2006)
4. Leskovec, J., Backstrom, L., Kleinberg, J.M.: Meme-tracking and the dynamics of the news cycle. In: SIGKDD 2009, pp. 457–466 (2009)
5. Kanhabua, N., Blanco, R., Matthews, M.: Ranking related news prediction. In: SIGIR 2011, pp. 755–764 (2011)
6. Pouliquen, B., Steinberger, R., Deguernel, O.: Story tracking: linking similar news over time and across languages. In: Colling 2008 MMIES Workshop, pp. 49–56 (2008)

7. Shan, D., Zhao, W.X., Chen, R., Shu, B., Wang, Z., Yao, J., Yan, H., Li, X.: Eventsearch: a system for event discovery and retrieval on multi-type historical data. In: Proceedings of the 18th ACM SIGKDD International Conference on Knowledge Discovery and Data Mining, pp. 1564–1567 (2012)

8. Leskovec, J., Grobelnik, M., Milic-Franling, N.: Learning sub-structures of document semantic graphs for document summarization. In: Proceedings of the 7th International Multi-Conference Information Society (2004)

9. Kalfoglou, Y., Kalfoglou, Y., Domingue, J., Domingue, J., Motta, E., Motta, E., Vargas-Vera, M., Vargas-vera, M., Shum, S.B., Shum, S.B.: myPlanet: an ontology-driven Web-based personalized news service. In: Proceedings of the IJCAI 2001 Workshop on Ontologies and Information Sharing (2001)

10. Hou, L., Li, J.Z., Tang, J., Liu, Y.K., Zheng, Q.: Newsminer: multifaceted news analysis for event search. To be Appeared in ACM Transactions on Information System 9(4) (2012)

11. Brank, J., Grobelnik, M., Mladenić, D.: A survey of ontology evaluation techniques. In: Proceedings of the Conference on Data Mining and Data Warehouses, SiKDD 2005 (2005)

12. Subhashini, R., Akilandeswari, J.: A survey on ontology construction methodologies. International Journal of Enterprise Computing and Business Systems 1(1), 60–72 (2011)

# Pharmaceutical Semantic Database Query Mechanism Based on KeyWords

Juan Sun[1,2], Jinguang Gu[1,2], and Zhisheng Huang[3]

[1] Computer Science and Technology, Wuhan University of Science
and Technology,Wuhan ,430065, P.R. China
[2] Hubei Province Key Laboratory of Intelligent Information Processing
and Realtime Industrial System, Wuhan, 430065, P.R. China
[3] Department of Computer Science, Vrije University of Amsterdam,
1081hv Amsterdam, The Netherlands

**Abstract.** With the development of the Internet, Semantic search technology has been applied to various fields. In the pharmaceutical field, because of the huge number and wide range of the drugs on the market, an effective semantic retrieval system can bring great convenience for doctors. However, for doctors who have no background knowledge of semantic technologies, to write formal semantics search statement is still very difficult. Therefore, this paper proposes a pharmaceutical semantic database query mechanism which based on keywords. In this query mechanism, by analysis of ontology database and establishment of Chinese - English translation mapping, we can Transform Chinese keywords into English keywords. Then match the English keywords with ontology database to build a list of triples. And finally, formed Corresponding SPARQL query by analyze the list.

**Keywords:** semantic retrieval, SPARQL, ontology database.

## 1 Introduction

From today to 1998 when the founder of the World Wide Web,Tim Berners-Lee[1],proposed the concept of the Semantic Web,it has been more than 10 years later. During this period Semantic Web related theories and technologies are developing rapidly, and become a hot topic. Currently, theres a large number of structured (XML, Web Database, etc.) or semi-structured information exist on the Internet, and with the passage of time, these information is still increasing. In order to provide a common semantic framework, the concept of Ontology has been introduced. Ontology is a conceptual model which could describe a system on sematic level. Its purpose is to share and reuse knowledge in different applications and organizations by describe knowledge in a general way to provide common understanding of the concept. RDF is a language which is used to describe the ontology, it stores information in the form of triples.

On the other hand, traditional technology of keyword search can not meet the growing needs of users while data on the internet becomes lager and more

G. Qi et al. (Eds.): CSWS 2013, CCIS 406, pp. 199–206, 2013.

complex. Semantic-based retrieval techniques [2] has become an urgent need. SPARQL[3] is a formal query language based on RDF document. Syntactically, SPARQL as a statement language, is similar to SQL database, the standard query language. It defines relationship between query contexts by using a certain pattern.

## 2   Related Research

Nowadays, pharmacy news resources are unusually large at home and abroad, however, World Wild Web provides us with a method to get and convey information quickly. Huge amount of pharmacy news is a challenge to the doctor's memory. Meanwhile, traditional retrieval system can't adequately express semantic information, which makes search results often unsatisfactory. Thus, a pharmaceutical information retrieval system based on the semantic processing [4] is an important problem urgently needed to solve at present.

In the field of semantic retrieval, researches of inquiring semantic information through keywords is not sufficient yet, many of which is still less mature. Currently, the research work of inquiring RDF data in the form of keywords can be divided into two types according to the difference of dispose pattern[5]:

- **(K-A):** The pattern of the query keyword directly structuring search result, such as BLINKS[6].
- **(K-Q-A):** The pattern that obtains search results by inquiring statement through inquiring the configurations of keywords like SemSearch[7] , SPARK[8] and so on.

BLINKS is a two-way indexing and query processing scheme, which is for searching top-k keywords in the map, its basic train of thought is that Dividing the map into blocks (subgraphs) more than one, building two-way indexing for each block to sum news of storage blocks, in addition, storing detailed block news in each block to speed up block search. On the background of mass information, BLINKS makes it possible to quckily obtain search results. But it is short in supporting users to input attribute name or relationship name as the query keyword.

SemSearch is a structurized keyword interface. By using certain defined frameworks and operators and it can help users to build up possible ontology queries so that the purpose of semantic search can be achieved. However SemSearch's keyword inquire can only be conducted in pre-defined pattern and operator. The semantic expression of SemSearch is limited and less flexible.

SPARK is a semantic retrieval system based on keywords, whose core idea is to utilize word - entity module to make the words that inputed by users mapped to the resources of the reality, then use the algorithm of query graph to match these resources into a query graph based on semantic connections, finally, the query ordering module utilizes sort algorithm to evaluate the query graph, and feedbacks the structured query statement which has been created and meets users' requirements to them. The efficiency of this algorithm is comparably high. Due to the coars-grained informtion after the querygraph's abstraction, however, some inquiries fail to get satisfying results.

# 3 Experimental Procedure

This article provides a way to transform Chinese keywords into SPARQL, using Drugbank[9] as ontology library which is a bioinformatics and chemical informatics database provided by University of Alberta. Drugbank combines detailed drug data (chemical, pharmacological and pharmaceutical) and comprehensive drug target information (sequence, structure and pathways). It contains 6729 entries drugs, 1,465 kinds of FDA (U.S. Food and Drug Administration - Food and Drug Administration) approved small molecule drugs, 132 FDA-approved biotech drugs (protein / peptide), as well as 86 kinds of nutrition and 5076 kinds of test drug.

Drugbank is a medication-centric RDF document, It stores drug-related information in triples. As shown in the following triples, it describes that the genericName of the drug is Lepirudin, this drug could be used for the treatment of heparin-induced thrombocytopenia.

```
<http://www4.wiwiss.fu-berlin.de/drugbank/resource/drugs/DB00001>
<http://www4.wiwiss.fu-berlin.de/drugbank/resource/drugbank/genericName>
"Lepirudin" .
<http://www4.wiwiss.fu-berlin.de/drugbank/resource/drugs/DB00001>
<http://www4.wiwiss.fu-berlin.de/drugbank/resource/drugbank/indication>
"For the treatment of heparin-induced thrombocytopenia" .
```

**Fig. 1.** An Example of triple

The convert method provided in this article is also a medication-centric process. It consists of three steps:

- Transform Chinese keywords into English keywords;
- Match the English keywords with drugbank to generate a triple map;

Analysis of the triple map to find the relationship between the keywords, and then generate SPARQL query.

## 3.1 Chinese keywords to English keywords

Since a good translation tools which could meet the requirements of semantic retrieval application dosen't exist as so far, this system will translate Chinese keywords into English keywords from the following two aspects:

- Use the exist data file from public platforms(such as DBpedia);
- Analysis of RDF document, extract the keywords from it and translated these words into English manually;

Divided the nodes in drugbank into the following categories:

**Table 1.** Ontology node type

| Type | Remark |
|---|---|
| concept_Drug | Pharmaceuticals node, refers to a certain kind of drugs. |
| concept_Text | Text node, typically describes attribute information of drugs such as state, indications, generic name etc. |
| concept_URI | Resource node, describes resources uri attribute , non-triple. |
| relation_Property | Attribute node, describes specific attributes of resources, Normally exist between resources node and text node. |
| relation_Node | Relationship node, Describe the relationship between nodes. |

DBpedia is an example of Semantic Web application . It extract structure information from Wikipedia to strengthen the Wikipedia search function , and links the other databases toWiki. At present , there are a big number of link library which linked other ontology libraries to Wikipedia published on DBpedia, such as drugbank_links.nt. It connect knowledge in drugbank and Wikipedia. The data format of this file is shown in the following triples:

```
<http://dbpedia.org/resource/Lepirudin>
<http://www.w3.org/2002/07/owl#sameAs>
<http://www4.wiwiss.fu-berlin.de/drugbank/resource/drugs/DB00001> .
<http://dbpedia.org/resource/Cetuximab>
<http://www.w3.org/2002/07/owl#sameAs>
<http://www4.wiwiss.fu-berlin.de/drugbank/resource/drugs/DB00002> .
```

**Fig. 2.** Data Format of drugbank_links.nt

Since the knowledge in drugbank_links.nt is described in English, it can't be used to translate keywords directly. So we introduced the drugbank-chinese.nt. This file is translated by a semantic research team of Wuhan University of Science and Technology. It contains all drugs Chinese name which appears in drugbank. The contents of this file is shown in Fig3.Combined these two files could translate Chinese keywords into English keywords, and mark such words with a label show in table 1.

Besides the drug names, theres a large number of drug-related knowledge exist in drugbank, such as pharmaceutical properties, classification and description and so on. First, analyzes the triples in drugbank, extract characteristic words from these triples according to theirs location, and then mark them with labels.

Then, translate the processed result into Chinese, an English word can be indicated by more than one Chinese word.Consolidated results of the above two steps to form a text file named: WordSets.txt to store data in certain format. The data format of this file is shown in Fig4.

<http://www4.wiwiss.fu-berlin.de/drugbank/resource/drugs/DB00001>
<http://ontoweb.wust.edu.cn/medicine#chineselabel>
"来匹卢定" .
<http://www4.wiwiss.fu-berlin.de/drugbank/resource/drugs/DB00002>
<http://ontoweb.wust.edu.cn/medicine#chineselabel>
"西妥昔单抗" .

**Fig. 3.** Data Format of drugbank-chinese.nt

来匹卢定/Lepirudin/CONCEPT_DRUG
血小板减少/thrombocytopenia/CONCEPT_TEXT
描述,描绘/description/RELATION_PROPERTY

**Fig. 4.** Example of WordSet

Here is a picture to show how did the WordSet.txt fromed:

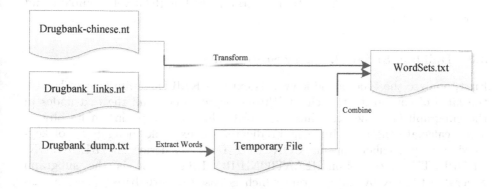

### 3.2   Matching Words and Spanning Subgraph

First, define drugbank as graph G, in which each triple as a node E, thus drug-bank could be expressed as: G = E1, E2, E3 ... En instead. Then, build an index file named index.txt to record the relationship between the nodes in graph G by program, the format of each record in index.txt is: [row number/pre-node row number/rear-node row number]. Definition 1: pre-node: Ex is the pre-node of Ei only if the subject of Ei is the object of Ex, which means Ei could be deduced from Ex . Definition 2: rear-node: Ex is the rear-node of Ei only if the object of Ei is the subject of Ex, which means Ex could be deduced from Ei. Here are a few examples:

**Input data:**
<http://www4.wiwiss.fu-berlin.de/drugbank/resource/drugs/DB00001>
<http://www4.wiwiss.fu-berlin.de/drugbank/resource/drugbank/
primaryAccessionNo>"DB00001" .
<http://www4.wiwiss.fu-berlin.de/drugbank/resource/drugs/DB00001>
<http://www.w3.org/1999/02/22-rdf-syntax-ns#type>
<http://www4.wiwiss.fu-berlin.de/drugbank/vocab/resource/class/Offer> .
<http://www4.wiwiss.fu-berlin.de/drugbank/vocab/resource/class/Offer>
<http://www.w3.org/2002/07/owl#equivalentClass>
<http://dbpedia.org/ontology/Drug> .
**Output data:** 1/0/0. 2/0/3. 3/2/0

After completion of the above preparations, translate Chinese keywords which inputed by users into English keywords through the mapping file. Define an empty subgraph Gs= . Then match the keywords with drugbank one by one . The set of English keywords could be expressed as K=kw1, kw2,..., kwn. If kwi matched with the node Ei, then add Ei into subgraph Gs. Until all the keywords exist in K are matched completly, then find all the pre-nodes and rear-nodes of the nodes exist in Gs through the index file above-mentioned, and add them into Gs as well. Thus, there forms with a subgraph which is Relatively complete and highly relevant with K.

## 3.3   Building SPARQL Query Statements

First, analyse the type of the keywords exist in K .If there are keywords with the label of concept_text, such as "thrombocytopenia", find the text-nodes in the subgraph Gs which contains the word "thrombocytopenia" in its object. Then calculate the most frequent predicate. For example, heres a set of keywords : K=[thrombocytopenia/CONCEPT_Text], [ genericName/RELATION_PROPERTY], [description/RELATION_PROPERTY], Gs is the subgraph generated from K by the program which is based on drugbank. There are 28 nodes contains the word "thrombocytopenia" in Gs, and 13 in them have the same predicate:

**<http://www4.wiwiss.fu-berlin.de/drugbank/resource/drugbank/tox
icity>**

Thus, it can get the following triple: (?s ,toxicity ,thrombocytopenia). On the other hand, Since the word "genericName" and "description" are used to describe the attribution of drugs, so we can get the following triple: (?s ,genericName ,?o1)(?s, description, ?o2). Merge these three triples. finally we get the following formal expression:

**?s,?o1,?o2 ← (?s ,toxicity ,thrombocytopenia)(?s ,genericName ,?o1)
(?s ,description ,?o2)**

# 4   Results

In order to verify the effectiveness of the mechanism, this paper selects a set of keywords as test data. figure.7 shown below lists the set of keywords and the corresponding formal expression:

**Table 2.** Result

| Chinese Keywords | Formal Expression |
|---|---|
| 来匹卢定 | ?o←(?s , genericName ,Lepirudin) ∩ (?s , description ,?o) |
| 来匹卢定 毒性 | ?o←(?s , genericName ,Lepirudin) ∩ (?s , toxicity ,?o) |
| 治疗 血小板减少 药物 | ?o←(?s , genericName ,?o) ∩ (?s ,indication ,thrombocytopenia) |
| 液态 药物 | ?o←(?s , genericName ,?o) ∩ (?s ,state ,Liquid) |

As we can see in Table2, this mechanism could converted Chinese keywords into certain formalized expression effectively. The formal expression could be converted into appropriate SPARQL query by the further process. For example:

$$?o \leftarrow (?s , genericName ,Lepirudin)(?s , toxicity ,?o)$$

This expression could be converted into the following SPARQL query:
PREFIX property:<http://www4.wiwiss.fu-berlin.de/drugbank/resource/drugbank/> Select ?o
From <drugbank.rdf>
Where{

?s property: genericName ?text.

?s property:toxicity ?o

FILTER regex(?text, Lepirudin, i).
}

# 5   Conclusion

Semantic retrieval pattern is a trend in the future search field, however, the formal semantic query language is not accepted by a large number of common users who have no background knowledge. Therefore, this article puts forward the way in which keyword could be turned into SPARQL inquiry, which effectively solved this contradiction. It enables common users not only to save their former search habits, but also to obtain comparably acurrate result data, which is of great significance to semantic research field. However, there are still many challenges in the process of transforming and dealing with Chinese keywords, the focus of work in the future will be on two aspects, on the one hand is the analysis of Chinese keywords, in order to tansform them into English keywords more accurately and more intelligently; on the other hand, to combine the user's usage habits with historical records input to generate SPARQL statements so as to make it obtain more accurate search results.

**Acknowledgement.** This work was partially supported by a grant from the NSF (Natural Science Foundation) of China under grant number 60803160, 61272110 and 61100133, the Key Projects of National Social Science Foundation of China under grant number 11ZD&189, and it was partially supported by a grant from NSF of Hubei Prov. of China under grant number 2009CDA136. It was partially supported by NSF of educational agency of Hubei Prov. under grant number Q20101110, and the State Key Lab of Software Engineering Open Foundation of Wuhan University under grant number SKLSE2012-09-07. It was partially supported by Program for Outstanding Young Science and Technology Innovation Teams in Higher Education Institutions of Hubei Province, China under grant number T201202.It was partially supported by key technologies r&d program of Wuhan under grant number 201110821236.

# References

1. Berners-Lee, T., Hendler, J., Lassila, O.: The semantic Web. Scientific American 284(5), 34–43 (2001)
2. Luan, Y., Ding, E., Luo, B.: Semantic Search Based on Ontology[J]. Computer Engineering and Applications 28(41), 156–159 (2005)
3. Pérez, J., Arenas, M., Gutierrez, C.: Semantics and Complexity of SPARQL. In: Cruz, I., Decker, S., Allemang, D., Preist, C., Schwabe, D., Mika, P., Uschold, M., Aroyo, L.M. (eds.) ISWC 2006. LNCS, vol. 4273, pp. 30–43. Springer, Heidelberg (2006)
4. Huang, Z., Zhong, N.: Massive semantic data processing. Higher Education Press, Beijing (2012)
5. Li, H., Qu, Y.: Keyword-based RDF data query method. Journal of Southeast University: Natural Science 40(2), 270–274 (2010)
6. He, H., Wang, H., Yang, J., Yu, P.: Blinks:ranked keyword searches on graphs. In: Proceedings of the 2007 ACM SIGMOD International Conference on Management of Data, pp. 305–316. ACM (2007)
7. Lei, Y., Victoria, U., Enrico, M.: Semsearch: a search engine for the semantic web, pp. 238–245 (2006)
8. Zhou, Q., Wang, C., Xiong, M., Wang, H.: SPARK:adapting keyword query to semantic search, pp. 694–707 (2008)
9. Huang, Z.: Interleaving Reasoning and Selection with Semantic Data. In: Proceedings of the 4th International Workshop on Ontology Dynamics (IWOD 2010), ISWC 2010 Workshop, pp. 122–129 (2010)

# Approach for Automatic Construction of Ontology Based on Medication Guide

Yuting Lu[1,2], Jinguang Gu[1,2], and Zhisheng Huang[3]

[1] Computer Science and Technology, Wuhan University of Science and Technology,
Wuhan 430065, China
[2] Hubei Province Key Laboratory of Intelligent Information Processing
and Real-time Industrial System, Wuhan 430065, China
[3] Department of Computer Science, Vrije University of Amsterdam,
1081hv Amsterdam, The Netherlands

**Abstract.** Medication guide ontology has great significance for semantic study of rational use of drugs. Nowadays, medical ontology is mainly
constructed manually, which spends a lot in time and manpower, and
is also difficult to ensure the accuracy. This paper proposes a method
of constructing the domain ontology of medication guide automatically.
The method uses semantic pattern analysis to extract information, opti-
mizes the extracted concepts and relations, generates medication guide
ontology, and then provides the interface of pattern expansion to adapt
to the changing medication guide data. Finally, it realizes the automatic
construction of ontology system, and conducts experiments on to verify
the feasibility of the system.

**Keywords:** Domain ontology, Ontology construction, Medication guide,
Rational use of drugs.

## 1 Introduction

Along with the reform of the medical care system and the increase of medical
disputes about the irrational drug use, rational drug use problem receives more
and more attention. And the growth of kinds of drugs is incurring unprecedented
pressure on clinician. At present, depending only on artificial intervention can't
meet all the requirements of the rational use of $drugs^{[1]}$. Therefore, more and
more semantic researches on rational use of drugs arise. Through the semantic
analysis of medication guide and patient data, it can automatically monitor
the rational use of drugs. The construction of medication guide ontology is an
important part of semantic study of rational use of $drugs^{[2]}$.

Nowadays, medical ontology is mainly constructed manually. However, if the
large amount data of medication guide completely rely on artificial construction,
it will consumes a lot of human resources and time, and is also difficult to guarantee the accuracy. This paper proposes an approach for automatic construction
of ontology based on medication guide to dig into medication guide and extract
ontology concepts as well as relations between them, so that it can generate
medication guide ontology quickly and accurately.

G. Qi et al. (Eds.): CSWS 2013, CCIS 406, pp. 207–214, 2013.
© Springer-Verlag Berlin Heidelberg 2013

## 2   Current Research

As the biomedical information updates constantly and rapidly, the construction of medication guide ontology is increasingly becoming a huge project. It needs domain expert and ontology engineer to work in cooperation. However, there is no mature theory to guide the ontology construction. Most of the construction methods have different processes, due to different bases of subject fields and specific projects.

Medical informatics experts have done some researches in the field of medication guide, including using ontology construction tools to artificially construct medication guide ontology, such as *Protégé*[3,4], and constructing medication guide ontology based on *thesaurus*[5].

The first method extracts a series of categories and properties by analyzing a large number of concepts of medical informatics, and uses ontology construction tools to annotate ontology. This method can receive more accurate properties of medication guide information because of a lot of domain experts' analyses, but it will consumes a lot of domain experts and their time owing to the huge data of medication guide. Moreover, manual annotation method is very difficult to maintain, as the growth of the biomedical information is more and more rapid.

The second method starts with the sentence segmentation of unstructured data of medication guide. Then it annotates ontology properties by using the sentence segmentation data to match the medical thesaurus. Thesaurus method avoids the consumption of a large number of human resource and time on the analysis of medical informatics concepts. However, it may produce unreasonable or ambiguous concepts. In addition, only using thesaurus to construct medication guide ontology may lead to extract relationship inaccurately. And it is difficult to receive a satisfying result when analyzing the complex sentences.

## 3   Construction Method

This paper proposes an approach for automatic construction of ontology based on medication guide by analyzing the characteristics of medication guide ontology, which is a domain-special ontology. This method can be divided into the following several parts according to the characteristics of the medication guide and different stages of the ontology construction: (as shown in Figure 1)

- The collection and pretreatment of medication guide. This step is to preliminarily sort out the original medication guide data, to remove the content that has nothing to do with the regulations of drug use, and to extract descriptions of the drug use regulations;
- Data segmentation of medication guide and relation extraction. This step is to further analyze the information which is processed in the step one through pattern segmentation and natural language processing, to extract the concepts and relations of drug use regulations, and to receive a series of n-tuples to form structured data of drug use regulations;

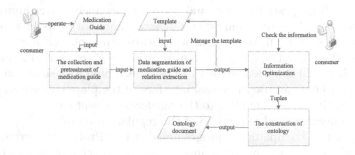

**Fig. 1.** The flow chart of medication guide ontology construction

- Information Optimization. This step continues to remove redundant data more deeply and proofread the information which is processed in step two in order to improve the accuracy of the concept extraction;
- The construction of ontology. This step is to build ontology document according to the data which is processed in the step three.

## 3.1 The Collection and Pretreatment of Medication Guide

The medication guide data can exist in different forms, such as document, book and so on. What's more, different document data also can have different forms. The information on a document is composed of unstructured natural language, thus we can't get the information of medication guide directly. Therefore, it needs to preliminarily sort out the original medication guide data firstly through human intervention in order to extract descriptions of drug use regulations and receive relatively structured natural language. The following is an example after structuring: These are extracted from antibiotics medication guide. This step provides data that the ontology construction need.

| The sentence after pretreatment |
| --- |
| 如果患者原发疾病可以治愈，那么抗菌药物用药方法为预防用药效果为可能有效； |
| 由真菌等病原微生物所致的感染亦有指征应用抗菌药物。 |

**Fig. 2.** The sentence after pretreatment

## 3.2 Data Segmentation of Medication Guide and Relation Extraction

Customers input the pre-processed data of medication guide into the system, then the system analyzes the data according to the templates. Each drug use regulation will be extracted and receive a n-tuple (Property, Object, Drug, Effect, Usage). The information received through extraction may be different according to the regulation template. According to the analysis of the medication guide, we can get an basic five-tuple, which represents the attribution of things, the clinical situation, the drug, pesticide effect and usage, respectively. After inputting the data, the system will analyze and extract in accordance with this basic template. If extraction is successful, the system will return a n-tuple. Otherwise, system will match other templates step by step by priority. If extraction is ultimately unsuccessful, it will send a feedback to customer about the situation. At this point, the customer can use the visual interface of template extensions to add a new template that can be applied to the unsuccessful sentence.

(1) The generation and realization of the regulation template

The regulation template is programmed with Prolog (Programming in Logic)[7,8] which is a logic programming language. The characteristics of Prolog are strong logic, brief grammar, and plainness. Programming the template with prolog can efficiently analyze regulations and extend templates easily.

The following is a semantic extraction regulation by prolog:

$$rule(Property, Object, Drug, Effect, Usage) \rightarrow cause(Property, Object), \\ separators, peroration(Drug, Effect, Usage). \tag{1}$$

$$cause(Property, Object) \rightarrow cHeader, property(Property), leadTo, \\ object(Object). \tag{2}$$

$$peroration(Drug, [], Usage) \rightarrow pHeader, drug(Drug), sp, usage(Usage). \tag{3}$$

The first regulation 'rule' is the extraction regulation for whole sentence; 'separators' is the separator of sentence(such as a comma); 'cause' is the extraction regulation of the first half sentence, including some important elements 'Property' and 'Object'; 'peroration' is the extraction regulation for the second half sentence, including the elements 'Drug', 'Effect', 'Usage'. The second and third regulations are the extraction regulation ('cause') of the first half sentence and the extraction regulation ('peroration') of the second half sentence respectively in detail.The following in Figure 3 is such an example of extraction. The result of extraction includes Property, Object, Drug, Effect and Usage. If some of elements don't exist in the sentence, this element is null. Such as the second example in Figure 3, it doesn't mention the pesticide effect, so 'Effect' is null. It is seen that if the sentence pattern or sentence expression changes, it only needs to add the regulation 'cause' or 'peroration' to meet the requirements of the change of the sentence without affecting the previous regulations. This makes it more convenient to extend the regulation template.

| The sentence before extraction | Property | Object | Drug | Effect | Usage |
|---|---|---|---|---|---|
| 如果患者原发疾病可以治疗，那么抗菌药物用药方法为预防用药效果为可能有效 | 原发疾病可以治疗 | 患者 | 抗菌药物 | 可能有效 | 预防用药 |
| 如果患者昏迷，那么建议常规抗菌药物用药方法为不用预防用药 | 昏迷 | 患者 | 抗菌药物 | | 不用预防用药 |

**Fig. 3.** The example of the extraction

(2) Calling the regulation template

Although prolog itself has the function of strong logical reasoning, it can't satisfy the need of designing perfect human-computer interface, so that the extraction data can't present to consume and maintain well. So this system combines prolog with .net, a programming language easily designs graphical interface. And it uses the SWI-Prolog interface to C# to implement the functions of the input of medication guide data, the import of regulation template and the present of n-tuple analyzed by regulation template.(as shown in Figure 4)

**Fig. 4.** The system interface of automatic construction of ontology

(3)The management of template

Meanwhile, the medication guide data will be continuously increased and updated. And the existing templates may not completely apply to the new guideline. So the system provides the function of managing the template, which is realized through programming with prolog and .net. Consume can use the visual interface to add or update the template, when the result of extraction is unsatisfying, such as mentioned in part one.

## 3.3   Data Optimization

Extracting the ontology concepts and relations according to regulation temples may exist the situation the extracted information is inaccurate. Thus, the system will further optimize this. It is mostly divided into two parts: removing redundant information and proofreading the extracted information;

(1) Removing redundant information

The medication guide data is composed of unstructured natural language, however natural language expression will lead to the situation that a semantic rule is repeatedly expressed by more than one sentence. So the extracted information may exist a number of redundant information. And this system is aimed at the situation to delete duplication.

According to the analysis of the domain-specific data of medication guide, we can get a mapping table of redundancy relations, which gathers some basic common redundant information mapping, such as it shows in Figure 5. In this step, the system will search for the generated n-tuple in the mapping table. Suppose there are two sets of information: the first set is $(A_1,A_2,A_3,\ldots,A_n)$, the second set is $(B_1,B_2,B_3,\ldots,B_n)$. If $A_1$, $B_1$; $A_2$, $B_2$; $A_3$, $B_3$;$\ldots$; $A_n$, $B_n$ are n sets of redundant data, the system will judge the first set $(A_1,A_2,A_3,\ldots,A_n)$ and the second set $(B_1,B_2,B_3,\ldots,B_n)$ are a set of redundancy. Then, system will delete one of these and show the processed information to the consumer.

**Fig. 5.** A mapping table of redundancy relations

(2) Proofreading the extracted information

After the medication guide data is extracted to a series of n-tuples, the system will show the extracted information to consumer in a list, as well as pointing out which regulation extraction is unsuccessful. Consumer can proofread and update the extracted information according to the information feedback of the system. Finally, consumer returns the proofread information to the system. In this way, it will make the accuracy of the extracted data higher.

## 3.4    The Construction of Ontology

The construction of ontology builds ontology document in the form of RDF according to n-tuples which is processed in the previous step. The existing database of medicine ontology, $DrugBank$[6], is a richly annotated resource that combines detailed drug data with comprehensive drug target and drug action information in the form of in the form of standard RDF triples. However, some drug description is not specific enough. Sometimes there are three or four sentences in one triple so that it may ignore the hidden information in the sentence. That is unfavorable for the semantic query of the rational use of drugs.

In this paper, medication guide ontology will be extracted to more detailed descriptions. The system extracts the medication guide data to a series of n-tuples by pattern analysis in order to describe the drug information more flexibly and detailedly. We can see the example in Figure 6. It is seen that the RDF data can describe the drug use regulation briefly and plain. According to the corresponding predicate description, we can quickly find the information of drug use regulation. Finally, system will generate the ontology document to consumer for downloading. It's convenient to the SPARQL query in drug use regulation of the rational use of drugs.

**Fig. 6.** The generated RDF data

# 4  Experiments

In the experiment, we choose the antibacterial drug medication guide as the experimental data in order to verify and evaluate the above approach for ontology construction. And we input this data into the system of automatic construction of ontology. Meanwhile we also manually construct the ontology, and contrast the two results. Then the result shows that the approach for automatic construction identifies 247 medication rules and 988 domain concepts of antibacterial drug medication guide, the approach for manual construction indentifies 260 medication rules and 1251 domain concepts of antibacterial drug medication guide. The automatic construction approach can extract 78% of the domain concepts. In addition, it can basically ensure the accuracy the consistency of extracting concepts in the logic. What's more, the information feedback module and the template extensions module will increase the accuracy of the extraction.

The experiment proves the feasibility of this method. Furthermore, limited to the current natural language processing technology, the pretreatment of medication guide of this method needs to manually complete. This still remains to improve and perfect.

# 5  Conclusions

The study of rational drug use is a hot issue, and the construction of ontology based on medication guide is the most basic part of this. This paper proposes an approach for automatic construction of ontology based on medication guide to extract the drug use regulations. It increases the flexibility of processing medication guide data, avoids the manual data cleansing of a lot of medication guide data, and shortens the ontology construction cycle. And it also adds the module of data optimization to increase the accuracy of the concept extraction. Finally, through the processing of the antibacterial drug medication guide, the feasibility and effectiveness of the method is verified. This has a significance to the semantic study of the rational use of drugs.

**Acknowledgement.** This work was partially supported by a grant from the NSF (Natural Science Foundation) of China under grant number 60803160, 61272110 and 61100133, the Key Projects of National Social Science Foundation of China under grant number 11ZD&189, and it was partially supported by a grant from NSF of Hubei Prov. of China under grant number 2009CDA136. It was partially supported by NSF of educational agency of Hubei Prov. under grant number Q20101110, and the State Key Lab of Software Engineering Open Foundation of Wuhan University under grant number SKLSE2012-09-07. It was partially supported by Program for Outstanding Young Science and Technology Innovation Teams in Higher Education Institutions of Hubei Province, China under grant number T201202.It was partially supported by key technologies r&d program of Wuhan under grant number 201110821236.

# References

1. Cui, R., Li, J.: Utilization Research of Medicament and Reasonable Utilization of Medicament. Hospital Administration Journal of Chinese People's Liberation Army 11(2), 147–148 (2004)
2. Huang, L.: Ontology and Its Application in Bioinformatics. Journal of Medical Intelligence 2, 81–84 (2006)
3. Li, Y., Zhang, Z.: Research about Ontology Building Method. Computer Knowledge and Technology 30(5), 129–132 (2008)
4. Cui, L., Zhao, P.: Developing a Practical Clinical Medication Ontology with Protege. New Technology of Library and Information Service 11, 77–80 (2006)
5. Liu, Y., Shui, Z., Zhou, Y., Wang, Z.: Research on Automatic Construction of Chinese Traditional Medicine. New Technology of Library and Information Service (5), 21–26 (2008)
6. Wishart, D.S.: DrugBank: A General Resource for Pharmaceutical and Pharmacological Research. Molecular and Cellular Pharmacology 2(1), 25–38 (2010)
7. Porto, A.: A structured alternative to Prolog with simple compositional semantics. Theory and Practice of Logic Programming 11, 611–627 (2011)
8. Wielemaker, J., Huang, Z., van der Meij, L.: SWI-Prolog and the Web. Theory and Practice of Logic Programming 8(3), 363–392 (2008)

# Evaluating Article Quality and Editor Reputation in Wikipedia

Yuqing Lu[1], Lei Zhang[2], and Juanzi Li[1]

[1] Shenzhen Key Laboratory of Broadband Network and Multimedia,
Graduate School at Shenzhen, Tsinghua University
[2] Department of Computer Science and Technology, Tsinghua University
{yuqing.lu,ljz}@keg.cs.tsinghua.edu.cn,
zhanglei@sz.tsinghua.edu.cn

**Abstract.** We study a novel problem of *quality and reputation evaluation* for Wikipedia articles. We propose a difficult and interesting question: How to generate reasonable article quality score and editor reputation in a framework at the same time? In this paper, We propose a dual wing factor graph(DWFG) model, which utilizes the mutual reinforcement between articles and editors to generate article quality and editor reputation. To learn the proposed factor graph model, we further design an efficient algorithm. We conduct experiments to validate the effectiveness of the proposed model. By leveraging the belief propagation between articles and editors, our approach obtains significant improvement over several alternative methods(SVM, LR, PR, CRF).

**Keywords:** factor graph, quality evaluation, editor reputation.

## 1 Introduction

Nowadays, the Wiki where every member can make contribution to its content is more and more popular. Take Wikipedia as an example, It has attracted 513 million page view per day in January 2012 and massive number of edits per day. The wiki pattern can fully embody the intelligence of the community. For example Wikipedia[1], WikiTravel[2], Neupedia[3], Hudong[4], they are all famous wiki websites where users can add, modify, or delete its content. Most wiki's content are created and edited collaboratively, any users can modify any articles based on his own cognition about the theme.

However, for most wiki websites, wiki articles quality are given by their assessment team. The work of assessment are all done manually. It's too time-consuming and boring. For example, in Wikipedia, there are 21,693,832 articles in 284 languages by April 2012. By June 2012, there are 3.29 million English articles have been assessed, while there are still 0.45 million English articles have never been assessed. What's more, these assessment teams can't assess quality for the articles immediately once they have been modified. Obviously article quality assessment is an urgent need.

---

[1] http://www.wikipedia.org/
[2] http://wikitravel.org/
[3] http://nunupedia.sourceforge.net/main.phtml
[4] http://www.hudong.com/

G. Qi et al. (Eds.): CSWS 2013, CCIS 406, pp. 215–227, 2013.

On the other hand, a wiki website is a collaborative website, users can be positive editors or vandalism. Assessing editor reputation is another obvious urgent job. It's also a time-consuming and boring job for wiki assessment team. Wikipedia assessment team can award some editors for their reputation. However, by now, there are still no explicit editor reputation in Wikipedia.

Considerable studies were conducted on wiki article quality. [2,3,4] try to detect vandalism based on content-persistence, [5,6,7] realize automatic vandalism detecting model based on NLP method, [9,10] propose a set of simple properties to help detect vandalism in Wikipedia. However, all these previous work are mainly presented to detect vandalism in Wikipedia. In this paper, we propose an interesting question: How can we generate reasonable article quality score and editor reputation score by considering both the informativeness of articles and editors? Our goal of this work is to generate article quality score and editor reputation in a unified framework. With the introduction of mutual reinforcement, our results can be more better than several baseline alternative methods. Obviously our problem is different from existed research. The task is no-trivial and poses a set of challenges:

- Assessing article's quality and assessing editor's reputation are two problems. How to solve these two difficult problems in a unified framework is a challenge.
- Different kinds of information can be used in the quality computation problem, such as article metadata, editor metadata, interaction between cooperative editors, interaction between editors and articles, interaction between articles which have been edited by some same editors, It's a non-trivial challenge to define a model that can incorporate all these information well.
- Wikipedia contains huge number of articles and editors. What's more, some articles may even have evolved from hundreds of versions. How to develop an effective and efficient algorithm that can deal with such a large number of data?

In this study, we try to conduct a systematic investigation of the problem of evaluating article quality and editor reputation in a unified framework. In order to address the above challenges we have discussed, we define the problem formally and define some important factors for article quality and editor reputation. Then we propose a dual wing factor graph model to compute article quality and editor reputation in a unified model. We further design an efficient algorithm to learn our proposed model. Our contribution include:

- We define article quality and author reputation in Wikipedia formally (Section 2) and quantitatively assess article quality score and author reputation based on our formulation.
- We unify the two difficult problems, assessment of articles and assessment of authors. Then we propose a unified factor graph model: Dual Wing Factor Graph(DWFG) to address the problem in Section 4.
- We conduct experiments to validate the effectiveness of our approach. We also compare our approach with some other baseline methods. Our approach clearly achieve better performance than other baseline methods (in Section 5).

## 2    Problem Formulation

In this section, we introduce some related concept for our problem definition, then we will define our problem formally. Firstly, we define the Wikipedia database as follows:

*DEFINITION 1.* For simplicity, **a Wiki database** can be viewed as a collection of collaboratively written articles. It can be represented as $D = \{d_i\}_{i=1}^N$ , where $d_i$ is an article in wiki database $D$ and $N$ is the amount of articles in the database.

Articles in wiki are written collaboratively. Most time, articles evolve by means of different editors' interaction together. Therefore, an article in wiki database $d_i \in D$ can be formally defined as a queue of version $d_i = \{v_{i1}, v_{i2}, \cdots, v_{im}\}$ , where $v_{ij}$ represents the $j$-th version of article $d_i$ and $m$ indicates the article $d_i$ contains $m$ versions altogether since it has been created.

Given a version $v_{ij}$ of article $d_i$, it can be represented as a four-tuple

$$(Text(v_{ij}), Time(v_{ij}), E(v_{ij}), C(v_{ij})) \tag{1}$$

where $Text(v_{ij})$ represents the text of this version, $Time(v_{ij})$ represents the modified time of this version, $E(v_{ij})$ represents this version's editor and $C(v_{ij})$ represents the comment of this version when it is committed.

*DEFINITION 2.* For each article $d_i$, we use **article quality** $q_i$ to represent its quality, $q_i = 1$ means good article and $q_i = 0$ means poor quality.

For each editor $e_i$, we use **editor reputation** $r_i$ to represent its reputation, $e_i = 0$ means high reputation editor and $e_i = 0$ means bad reputation editor.

In Wikipedia, its assessment work are mainly performed by members of WikiProjects. By WikiProject article quality grading scheme, articles can be graded as Featured Article, A-class, good article, B-class, C-class, start or stub. Featured article is the best article grade and stub class only contains very basic description of the topic.

*Problem*: How to calculate article quality and editor reputation in a united framework simultaneously?

## 3    Experiment Pre-processing

**Data collection**: We download English Wikipedia dataset at Wikipedia dumps website[5]. Via the downloaded dataset, we can extract all quality label for each labeled article. As shown in Table 1, it describes article quality distribution in our Wikipedia dataset. We can easily find that articles rated as FA, A, GA or B are much less then articles rated as C, Start or Stub, actually, the sum of articles rated as FA, A, GA and B only take up 3.184% in all. We label those FA, A, GA and B articles as high quality articles, i.e. $q_i = 1$. Besides these article label, We also extract rewards awarded to all editors. We label those editors who have been awarded as high reputation editors, i.e. $r_i = 1$.

**Feature Selection**: Refer to existed work [12,13], we choose many article features and editor features. Table 2 describes all features we have extracted for articles and editors respectively.

---

[5] http://dumps.wikimedia.org/enwiki/20120403/

**Table 1.** Article Quality Distribution

| FA | A | GA | B |
|---|---|---|---|
| 4,118 | 920 | 16,094 | 81,514 |
| C | Start | Stub | |
| 125,440 | 859,345 | 2,136,275 | |

We should note that, we use the 10-th feature $Iq_{ij}$ to represent the quality comparison result between article $a_i$ and $a_j$. If article $a_i$ has more than 6 features are better than article $a_j$, we empirically think article $a_i$ are better than $a_j$, we set $Iq_{ij} = 1$, otherwise, $Iq_{ij} = 0$.

We also use the 14-th feature $Ir_{ij}$ to represent the reputation comparison result between editor $e_i$ and $e_j$. If editor $e_i$ has more than 2 reputation features are better than editor $e_j$, we empirically think editor $e_i$ should have higher reputation than $e_j$, we set $Ir_{ij} = 1$, otherwise, $Ir_{ij} = 0$.

For each article to editor edge, if and only the article is a good quality article $q_i = 1$ and editor is a low reputation editor $e_j = 0$, we set $Iqr(ij) = 1$, otherwise $Iqr_{ij} = 0$.

**Table 2.** Feature List

| | No | Description |
|---|---|---|
| article | 1 | the length of article |
| quality | 2 | the number of image in article |
| features | 3 | the number of article's inlink |
| | 4 | the number of article's unique editor |
| | 5 | the number of positive editor |
| | 6 | if the article contains infobox |
| | 7 | the lifetime of article |
| | 8 | the lifetime of the current version |
| | 9 | the number of article version |
| article-article | 10 | the article quality comparison indicator |
| editor | 11 | the lifetime of the registered ID |
| reputation | 12 | the rate of positive edition |
| features | 13 | the number of positive edition |
| editor-editor | 14 | the editor reputation comparison indicator |
| article-editor | 15 | the article to editor feature indicator |

## 4   Our Proposed Approach

Shown as Fig. 1.a, it describes the original problem. Only considering about editor attribute, we can get Editor Reputation Graph, as shown in Fig. 1.b. Only considering about article quality, we can get Article Quality Graph, as shown in Fig. 1.c. In our work, we define a Dual Wing Factor Graph(DWFG), as show in Fig. 1.d. Besides these factors in Editor Reputation Graph and factors in Article Quality Graph, In the DWFG model, we utilize the interaction between editors and articles, we unify these two difficult problems into a unified framework by these inter-domain factor.

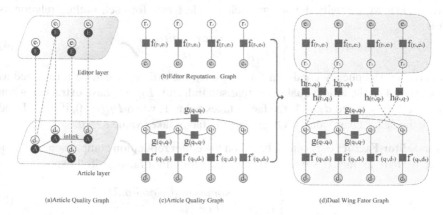

(a)Article Quality Graph          (c)Article Quality Graph          (d)Dual Wing Fator Graph

**Fig. 1.** An Example of DWFG model for evaluating article quality and editor reputation. In (b), (c) and (d), each gray circle $e_i$ represents an editor, and the value of white circle $r_i$ indicates whether the editor $e_i$ is an edito r with high reputation; each gray circle $d_i$ represents an article, and the value of white circle $q_i$ indicates whether the article $d_i$ is an article with high quality; each black square represents a factor in the DWFG model

## 4.1   Dual Wing Factor Graph

Based on all these feature we extract from our data set as discussed in Section 3, we de fine four types of factors in DWFG model: *local attribute factor*, *article-article factor*, *editor-editor factor* and *article-editor factor*.

**Local Attribute Factor:** We can evaluate article's quality based on article's local attributes we have selected. For each feature, we introduce a weight variable $\lambda_c$ for each feature $c$ to indicate its importance, then we can define the local attribute factor by the local entropy formally as follow:

$$f_c(q_i|d_{ic}) = exp(\lambda_c \cdot d_{ic} \cdot q_i) \tag{2}$$

where $d_{ic}$ is the value of the c-th feature for article $d_i$ and $q_i$ is the quality value of this article.

$$f_c'(r_i|e_{ic}) = exp(\lambda_c' \cdot e_{ic} \cdot r_i) \tag{3}$$

where $e_{ic}$ is the value of the c-th feature for editor $e_i$ and $r_i$ is the reputation value of this editor. In our work, we use $\lambda$ and $\lambda'$ to distinguish article's local factor and editor's local factor.

**Article-Article Factor:** We can make use of the outlink relation to make restriction article quality score. For simplicity, if an article $a_i$ has an outlink linking to another article $a_j$, then we can regard these two articles as similar articles under the same quality criteria. Then we formally define an article-article factor for each outlink relation as follow:

$$g(q_i, q_j) = \begin{cases} 1 & some\ condition\ holds \\ exp\ \mu & otherwise \end{cases} \tag{4}$$

where the dependency condition for article $a_i$ and article $a_j$ can be formalized as follow: Based on the article quality comparison indicator $Iq_{ij}$ we have extracted, when $Iq_{ij} = 1$, if $d_i = 0$ and $d_j = 1$, the factor takes value 1; when $Iq_{ij} = 0$, if $d_i = 1$ and $d_j = 0$, the factor takes value 1. Otherwise, the factor takes value $exp\ \mu$.

**Editor-Editor Factor:** For editors, based on the editor reputation comparison indicator feature we have extracted, we formally define editor-editor factor as follow:

$$g'(r_i, r_j) = \begin{cases} 1 & some\ condition\ holds \\ exp\ \nu & otherwise \end{cases} \tag{5}$$

where the dependency condition for editor $e_i$ and editor $e_j$ can be formalized as follow: Based on the editor reputation comparison indicator $Ir_{ij}$ we have extracted, when $Ir_{ij} = 1$, if $r_i = 0$ and $r_j = 1$, the factor takes value 1; when $Ir_{ij} = 0$, if $r_i = 1$ and $r_j = 0$, the factor takes value 1. Otherwise, the factor takes value $exp\ \nu$.

**Article-Editor Factor:** For article to editor edges, we make a simple and intuitive assumption: Editors with low reputation are more likely to make vandalism action on articles with high quality. Based on the assumption, we define formally the article-editor factor:

$$h(q_i, r_j) = \begin{cases} exp\ \xi & some\ condition\ holds \\ 1 & otherwise \end{cases} \tag{6}$$

where the dependency condition is defined based on our assumption, if and only if $q_i = 1$ and $r_j = 0$, the article-editor factor takes value $exp\ \xi$, otherwise it takes value 1;

## 4.2  Model Learning

**Objective Function:** Finally, we can define model objective function as the normalized product of equation 2-6 for all instances. In our work, we use $\Theta$ to denote the union collection of all weight, i.e., $\Theta = \{\lambda_c\}_c \cup \{\mu\} \cup \{\nu_c\}_c$ Finally, we can utilize Eqs 2-6 to define our object function for our dual wing factor graph.

$$p(Q, R|D, E, \Theta) = \frac{1}{Z} \prod_{q_i, q_j \in Q} \prod_{r_i, r_j \in R} \prod_{c \in C}$$
$$f_c(q_i|d_{ic}) \cdot f'_c(r_j|e_{jc}) \cdot g(q_i, q_j) \cdot g'(r_i, r_j) \cdot h(q_i, r_j) \tag{7}$$

where $Z$ is the normalized value, which sums up the conditional likelihood $P(Y, Z|E, D, \Theta)$ over all the possible labels of all the instances, and $Q$ and $R$ is the article quality value set and editor reputation set respectively.

We need to estimate weights for our model. First, in training procedure, we estimate these weights $\Theta$ by using maximum likelihood on training data.

$$p(\Theta|Q, R, D, E) = \frac{1}{Z} \prod_{q_i,q_j \in Q} \prod_{r_i,r_j \in R} \prod_{c \in C}$$
$$f_c(q_i|d_{ic}) \cdot f'_c(r_j|e_{jc}) \cdot g(q_i, q_j) \cdot g'(r_i, r_j) \cdot h(q_i, r_j) \quad (8)$$

**Learning Method:** We use L-BFGS, a quasi-Newton method for solving the nonlinear optimization problem(i.e. Eq.8). We also add a penalty term $\frac{1}{2} \|\Theta^2\| / \sigma^2$ into the objective function to avoid overfitting. It's a regularization method commonly used in maximum entropy and conditional random fields. With the leaned parameter $\Theta$, we can calculate article quality value and editor reputation, which are also a similar max-product inference algorithm. We will present our max-product inference algorithm in the next subsection.

We should note that, our proposed dual wing factor graph model can be viewed as a model generating from existing model. In Eq.8, if we fix weight $\xi = 0$, then the DWFG model can be viewed as a simple CRF model on article quality and editor reputation respectively. If we fix $\mu = 0$ and $\xi = 0$, only considering article quality, the model can be viewed as model like [9, 10], which only consider about article's local attributes. It will degenerate into a logistic regression classifier or a SVM classifier. Making all these parameter fixed as 0, except for $\lambda_2$, that is to say, if we only utilize article citation to estimate article quality, then our model will degenerate into a model like in [14].

## 4.3   Model Inference

Having learned the parameter $\Theta$, we can evaluate article quality and editor reputation in our DWFG model by maximum Eq. 7. We will first introduce the general algorithm to make inference in factor graph in brief and then given our algorithm.

Sum-product[17], based on message propagation, is a general algorithm to compute marginal function. It performs exact inference on a factor graph without cycles. In the sum-product algorithm, the marginal functions of a variable are passed between neighboring variable node and function node as belief messages. All messages begin propagation from all these leaves. Each nodes in the factor graph waits for message from its neighbor, once all its neighbors except for left neighbor have transmit there message to the node, it can transmit its message to the left neighbor and wait for the message respond from the neighbor. The process terminates when two messages have passed on every edge. At each node, the product of all incoming message, is its marginal function. In acyclic graph, the sum-product algorithm can get exact inference.

However, in our DWFG model, there exist cycles, we can't utilize the sum-product algorithm, then we propose an inference approach based on the loopy sum-product or max-sum[19].

To achieve an approximate inference for evaluating article quality and editor reputation, the algorithm contains multiple iterations for updating beliefs. Each iteration contains two stages: 1) all variable nodes transmit belief to their neighbor factor nodes simultaneously; 2) all factor nodes transmit belief to their related neighbor variable nodes. Here we denote the messages for delivering belief between variable nodes and factor nodes by $m_{x \to f}$ and $m_{f \to x}$. The symbol $m_{x \to f}$ represents the message sent to factor node from variable node and the symbol $m_{f \to x}$ represents the message sent to variable node from factor node. The messages can be formulated as follows:

$$m_{x \to f} = \sum_{h \in N(x)} m_{h \to x} \tag{9}$$

$$m_{f \to x} = \max_{\sim \{x\}} \{ f + \sum_{x' \in n(f) \backslash \{x\}} m_{x' \to f} \} \tag{10}$$

where the notation $\sum_{h \in N(x)} m_{h \to x}$ denote the sum of all belief having received from all neighbour factor nodes and the notation $\sum_{x' \in n(f) \backslash \{x\}} m_{x' \to f}$ denote the sum of all belief having received from all neighbor variable nodes except for $x$.

For each article, we can obtain the label for article quality or editor reputation via the belief messages calculated in the two inference stages in the last iteration as follows:

$$y_i = \begin{cases} 1 & if\ m_{x \to f} + m_{f \to x} > 0\ for\ some\ f \\ 0 & otherwise \end{cases} \tag{11}$$

Details on inference procedure is given in Algorithm 1.

---

**Algorithm 1.** Detailed algorithm for inference

---

**Require:**
All local attributes set $C$; The set of article-article relation $G$; The editor-editor relation set $G'$; The article-editor relation set $H$; The set of factor weight parameter $\Theta$ and number of iterations $I$

**Ensure:**
All articles quality value and editors reputation value
// update messages
1. **for** $i \leftarrow 1\ to\ I$ **do**
2.    update message $m_{x \to f}$ according to Eq. 9;
3.    update message $m_{f \to x}$ according to Eq. 10;
4. **end for**
// obtain article label
5. calculate $y_i$ for each article according to Eq. 11
// obtain editor label
6. calculate $y_i$ for each editor according to Eq. 11

---

## 5    Experiments

In this section, we will evaluate our method with some labeled dataset. We firstly introduce our labeled dataset and then describe some baseline methods which we choose to compare with our method. Finally, we make detailed analysis on the experiment result on our method and other baseline method.

**Table 3.** Detail on test dataset

| Dataset name | Category | Article size | Editor size |
|---|---|---|---|
| Computer science | Computer science | 4129 | 337764 |
| Physic | Physic | 2428 | 180764 |
| Chemical | Chemical | 1462 | 152174 |

## 5.1 Experiment Setting

**Dataset.** Since there is rarely previous work study the Wikipedia article quality value evaluation and editor reputation value evaluation, to the best of our knowledge, no existing benchmark dataset can be utilized for the evaluation of our experiment. We download data from http://dumps.wikimedia.org/enwiki/20120403/ and extract data from dump data to build our test dataset.

As each article has been classified as several categories, Based on article's category classification, we can collect all articles have been classified as the same category to build our test set. For simplicity, randomly, we collect all articles classified as "computer science" and then we collect all authors who have edited these articles belong to "computer science", we get a test dataset named "computer science". We also collect some other test datasets named "Physic", "Chemical" and so on. For the sake of brevity, we only show the results of datasets "computer science", "Physic" and "Chemical". Details on these three datasets are given in the Table 3.

**Baseline Methods.** We define several methods as baseline methods to compare assessment results with our approach, including Support Vector Machine (SVM), Logistic Regression Classifiers (LR), PageRank (PR), Conditional Random Field (CRF). Details about these algorithms have been omitted to save space.

**Evaluation Method.** We evaluate each method by $F_1$ Measure and Accuracy. Detailed comparison will be given later.

## 5.2 Results and Analysis

**Comparison Results.** We conduct our experiment in the 10-fold cross validation procedure. The experiments results comparison for articles and editors are shown in Table 4 and Table 5 respectively. In both table, the best performances have been highlighted in bold.

**Factor Contribution Analysis.** We further analyze the contribution or importance of each factor. We show the estimated weights for article local factors $\lambda_1,..., \lambda_9$ on three domains respectively and their average weight value in Table 6. We also show the estimated weights for editor local factors $\lambda'_{11}$, $\lambda'_{12}$ and $\lambda'_{13}$ on three domains respectively and their average weight in Table 7.

**Table 4.** Experiment results comparison on article dataset

|       |      | Computer | Physic | Chemical | All |
|-------|------|----------|--------|----------|-----|
| $F_1$ | SVM  | 0.215 | 0.298 | 0.468 | 0.246 |
|       | LR   | 0.243 | 0.303 | 0.475 | 0.265 |
|       | PR   | 0.205 | 0.287 | 0.455 | 0.239 |
|       | CRF  | 0.288 | **0.335** | 0.505 | 0.317 |
|       | DWFG | **0.313** | 0.329 | **0.552** | **0.335** |
| A     | SVM  | 0.842 | 0.820 | 0.815 | 0.830 |
|       | LR   | 0.831 | 0.847 | 0.803 | 0.831 |
|       | PR   | 0.825 | 0.816 | 0.786 | 0.815 |
|       | CRF  | 0.906 | 0.876 | 0.860 | 0.888 |
|       | DWFG | **0.925** | **0.912** | **0.893** | **0.915** |

**Table 5.** Experiment results comparison on editor dataset

|       |      | Computer | Physic | Chemical | All |
|-------|------|----------|--------|----------|-----|
| $F_1$ | SVM  | 0.549 | 0.543 | 0.526 | 0.543 |
|       | LR   | 0.551 | 0.538 | 0.556 | 0.549 |
|       | PR   | 0.528 | 0.517 | 0.543 | 0.529 |
|       | CRF  | 0.532 | 0.559 | **0.613** | 0.568 |
|       | DWFG | **0.568** | **0.606** | 0.589 | **0.585** |
| A     | SVM  | 0.887 | 0.864 | 0.895 | 0.883 |
|       | LR   | 0.885 | 0.874 | 0.893 | 0.884 |
|       | PR   | 0.825 | 0.816 | 0.786 | 0.814 |
|       | CRF  | 0.914 | 0.923 | 0.886 | 0.910 |
|       | DWFG | **0.956** | **0.943** | **0.962** | **0.954** |

**Table 6.** Article local factor parameter estimation results

|                | Computer | Physic | Chemical | Average |
|----------------|----------|--------|----------|---------|
| $\lambda_1$ | 3.833 | 4.121 | 5.677 | **4.544** |
| $\lambda_2$ | 5.11E-04 | 7.92E-04 | 2.39E-04 | 5.140E-04 |
| $\lambda_3$ | 0.141 | 0.059 | 0.064 | 0.088 |
| $\lambda_4$ | 2.031 | 1.732 | 1.633 | 1.798 |
| $\lambda_5$ | 1.253 | 1.021 | 0.878 | 1.051 |
| $\lambda_6$ | 1.77E-07 | 1.77E-07 | 1.94E-08 | 1.44E-07 |
| $\lambda_7$ | -4.650 | -4.650 | -4.429 | **-4.577** |
| $\lambda_8$ | -0.356 | -0.419 | -0.271 | -0.349 |
| $\lambda_9$ | 13.30 | 10.33 | 11.29 | **11.64** |

From Table 6, we can find that most of the article local factors make positive contributions to our task. Among all these feature, Feature 1 (the length of article), Feature 7 (the lifetime of article) and Feature 9 (the number of article version) which have been highlighted, are the most important features to identify whether an article is a good authoritative article. From Table 7 we can find that both Feature 12(the rate of positive edition) and Feature 13(the number of positive edition) are important features to

Table 7. Editor local factor parameter estimation results

| | Computer | Physic | Chemical | Average |
|---|---|---|---|---|
| $\lambda'_{11}$ | 0.037 | 0.063 | 0.056 | 0.052 |
| $\lambda'_{12}$ | 5.683 | 5.792 | 5.923 | **5.800** |
| $\lambda'_{13}$ | 7.992 | 7.531 | 8.247 | **7.923** |

identify whether an editor is a good authoritative editor. Feature 11(the lifetime of registered ID) is not important for editor reputation.

# 6  Related Work

Wiki article quality have been widely studied since these years and many approach have been developed to assess collaboratively edited articles. The main approach for assessing article quality can be classified into four categories [16]: Content-persist, NLP, Metadata-based, Citation-based. In this section, we will overview the main methods to evaluate article quality from existed literature.

**Content-Persist:** As detailed by Adler [5][4] and Zeng [7], text fragments speaks for its editor's reputation. If some text remains, its editor gain reputation.

**NLP:** Wang [9], K Smcts [10] and [11] utilize NLP-based features to analyze the quality of articles.

**Metadata-Based:** Stvilia [13] and West [8] measure the quality of articles in the Wikipedia by using metadata properties.

**Citation-Based:** McGuinness [14] propose a link-ratio algorithm similar with other citation-based algorithms to evaluate article quality.

In our work, we integrate metadata-based and citation-based approaches into the same model. We select many good features as the local attribute feature of article's quality and editor's reputation. What's more, to the best of our knowledge, distinct from these existed approaches, we can evaluate the article quality and editor reputation in a united framework simultaneously.

# 7  Conclusion

In this paper, we propose a new problem, predicting article quality and editor reputation simultaneously. We define a large number of features to describe reviews and reviewers respectively. Distinct from previous work, we also define a set of features: feature between articles, feature between editors and feature between article and editor. To incorporate all features we have defined, we propose a Review Feature Graph model. We further design an efficient max-sum algorithm which utilize belief propagation to perform model learning and inference. We also conduct experiments to compare our method with other baseline methods.

With some information about editor helping to evaluate article quality, we can evaluate article more accurately; With some information about article helping to evaluate editor reputation, we can evaluate editor more accurately. With the mutual enhancement effect, our approach achieve significant improvement overall other baseline methods.

Finally based on our learned weights, we analyse the contribution of each factor to evaluate article quality and editor reputation.

**Acknowledgments.** The work is supported by the Natural Science Foundation of China (No. 61035004, No. 60973102), 863 High Technology Program (2011AA01A207), European Union 7th framework project FP7-288342, and THU-NUS NExT Co-Lab

# References

1. Smith, T.F., Waterman, M.S.: Identification of Common Molecular Subsequences. J. Mol. Biol. 147, 195–197 (1981)
2. May, P., Ehrlich, H.-C., Steinke, T.: ZIB Structure Prediction Pipeline: Composing a Complex Biological Workflow Through Web Services. In: Nagel, W.E., Walter, W.V., Lehner, W. (eds.) Euro-Par 2006. LNCS, vol. 4128, pp. 1148–1158. Springer, Heidelberg (2006)
3. Foster, I., Kesselman, C.: The Grid: Blueprint for a New Computing Infrastructure. Morgan Kaufmann, San Francisco (1999)
4. Adler, B.T., Chatterjee, K., De Alfaro, L., Faella, M., Pye, I., Raman, V.: Assigning trust to Wikipedia content. In: Proceedings of the 4th International Symposium on Wikis. ACM Press (2008)
5. Adler, B.T., de Alfaro, L.: A Content-Driven Reputation System for the Wikipedia. ACM Press (2007)
6. Wu, Q., Irani, D., Pu, C., Ramaswamy, L.: Elusive vandalism detection in wikipedia: a text stability-based approach. In: Proceedings of the 19th ACM International Conference on Information and Knowledge Management, pp. 1797–1800. ACM Press (2010)
7. Zeng, H., Alhossaini, M.A., Ding, L., Fikes, R., McGuinness, D.L.: Computing trust from revision history. In: Proceedings of the 2006 International Conference on Privacy, Security and Trust: Bridge the Gap Between PST Technologies and Business Services, vol. 8, ACM Press (2006)
8. West, A.G., Kannan, S., Lee, I.: Detecting Wikipedia vandalism via spatio-temporal analysis of revision metadata. In: Proceedings of the Third European Workshop on System Security, pp. 22–28. ACM Press (2010)
9. Wang, W.Y., McKeown, K.R.: Got you!: automatic vandalism detection in Wikipedia with web-based shallow syntactic-semantic modeling. In: Proceedings of the 23rd International Conference on Computational Linguistics, pp. 1146–1154. Association for Computational Linguistic (2010)
10. Smets, K., Goethals, B., Verdonk, B.: Automatic vandalism detection in Wikipedia: Towards a machine learning approach. In: AAAI Workshop on Wikipedia and Artificial Intelligence: An Evolving Synergy, pp. 43–48. ACM Press (2008)
11. Itakura, K.Y., Clarke, C.L.A.: Using dynamic markov compression to detect vandalism in the wikipedia. In: Proceedings of the 32nd International ACM SIGIR Conference on Research and Development in Information Retrieval, pp. 822–823. ACM Press (2009)
12. Rassbach, L., Pincock, T., Mingus, B.: Exploring the Feasibility of Automatically Rating Online Article Quality (2008)

13. Stvilia, B., Twidale, M.B., Smith, L.C., Gasser, L.: Assessing information quality of a community-based encyclopedia. In: Proceedings of the International Conference on Information Quality, vol. 11. Citeseer (2005)
14. McGuinness, D.L., Zeng, H., Da Silva, P.P., Ding, L., Narayanan, D., Bhaowal, M.: Investigations into trust for collaborative information repositories: A wikipedia case study. In: Proceedings of the Workshop on Models of Trust for the Web. Citeseer (2006)
15. Page, L., Brin, S., Motwani, R., Winograd, T.: The PageRank citation ranking: Bringing order to the web. Stanford InfoLab (1999)
16. West, A.G., Chang, J., Venkatasubramanian, K.K., Lee, I.: Trust in collaborative web applications. In: Future Generation Computer Systems. Elsevier (2011)
17. Kschischang, F.R., Frey, B.J., Loeliger, H.A.: Factor graphs and the sum-product algorithm. IEEE Transactions on Information Theory 47, 498–519 (2001)
18. Loeliger, H.A.: An introduction to factor graphs. IEEE Signal Processing Magazine 21, 28–41 (2004)
19. Murphy, K.P., Weiss, Y., Jordan, M.I.: Loopy belief propagation for approximate inference: An empirical study. In: Proceedings of the Fifteenth Conference on Uncertainty in Artificial Intelligence, pp. 467–475. Morgan Kaufmann Publishers Inc. (1999)
20. Yang, Z., Cai, K., Tang, J., Zhang, L., Su, Z., Li, J.: Social context summarization. In: Proceedings of the 34th ACM SIGIR Conference (2011)
21. West, A.G.: Calculating and Presenting Trust in Collaborative Content. University of Pennsylvania (2010)
22. Blumenstock, J.E.: Size matters: word count as a measure of quality on wikipedia. In: Proceedings of the 17th International Conference on World Wide Web, pp. 1095–1096. ACM (2008)
23. Chin, S.C., Street, W.N., Srinivasan, P., Eichmann, D.: Detecting Wikipedia vandalism with active learning and statistical language models. In: Proceedings of the 4th Workshop on Information Credibility, pp. 3–10. ACM (2010)

# Ontology Construction in Tea Pest Domain

Jing Sun, Shaowen Li, Li Zhang, Chao Liu, Huiyuan Zhao, and Jingui Yang

[1] School of Information & Computer Science, Anhui Agricultural University, Hefei, P.R. China
[2] Anhui Provincial Key Laboratory of Agricultural Informatics, Hefei, P.R. China
{1943899358,514409760,461574525,770154836}@qq.com,
shwli@ahau.edu.cn, abd20000@163.com

**Abstract.** Ontology is a conceptual model for the expression of shared knowledge. In this paper, we propose a method to construct an ontology in the tea pest domain by ontology reasoning and evaluation. The constructed ontology lays a solid foundation for agricultural ontology services, such as diagnosis and control of tea pests, information resources sharing for tea farmers and knowledge reuse in the domain of tea science.

**Keywords:** agricultural ontology, ontology modeling, tea pest, OWL2.

## 1 Introduction

As we all know, there are more than 800 kinds of tea pest in China. The dangers of tea pests in tea production process have been a more prominent issue, seriously affecting the tea production. In order to effectively diagnose and control tea pests, scholars at home and abroad have carried out many different aspects of studies and explorations. Among them, agricultural knowledge service, a combination of information technology is an effective method to guide the agricultural production, to solve the production problems and to improve the quality of production.

At the same time, pest control is a wide-ranging and cross-disciplinary subject including lots of concepts and relationships. Therefore, this paper introduces ontology into agricultural knowledge service and constructs a tea pest domain ontology, which can express tea pest domain knowledge clearly and explicitly, realize knowledge sharing and reuse, and provide the foundation for ontology-based services, such as intelligent retrieval and diagnosis.

## 2 Agricultural Ontology

Ontology is a formal, explicit specification of a shared conceptualization [1]. It is composed of five basic modeling primitives, containing concepts, relationships, instances, axioms and functions. Agricultural ontology is defined as a set of formal, explicit specification of a shared conceptualization in agricultural science [2]. The construction procedures for agricultural ontology are shown in Figure 1.

G. Qi et al. (Eds.): CSWS 2013, CCIS 406, pp. 228–234, 2013.

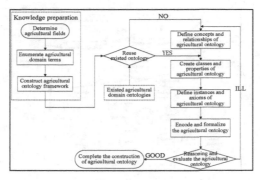

**Fig. 1.** Construction procedures for agricultural ontology

In the field of agricultural ontology, a lot of work has been carried out by many research institutions. For example, the Food and Agriculture Organization of the United Nations has built food safety ontology, fishery ontology, crop-pest ontology, etc [3-4]. These ontologies not only provide new ideas and methods for knowledge representation and management, but also effectively improve the promotion and application of agricultural knowledge service in practical production.

## 3  Tea Pest Domain Ontology Construction

### 3.1  Knowledge Acquisition

Cooperating with key laboratory of tea biochemistry and biotechnology in Anhui Agricultural University, we acquire and integrate tea peat domain knowledge mainly from books, expert experience and network. There are more than 400 common tea pests in China, and we classify the concepts based on biological taxonomy to ensure the ontology structure is clear and explicit.

During the process of knowledge acquisition, firstly we classify the concepts of tea pests (see Figure 2); then we build the relationships among concepts, and describe the particular characteristics of each common tea pest. In brief, the major taxonomic and

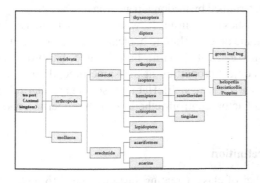

**Fig. 2.** Classification of concepts in tea pest domain ontology

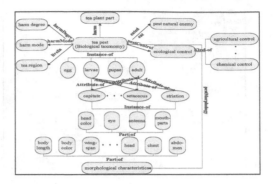

**Fig. 3.** Relationships among concepts in tea pest domain ontology

non-taxonomic relationships in tea pest domain ontology are shown in Figure 3. After organizing the concepts and relationships clearly, we generate the knowledge framework of tea pests, which will lay the foundation for the next steps of ontology construction.

### 3.2    Class Definition

To ensure the completeness and accuracy of domain concepts, ontology uses class to represent a set of individuals with common characteristics, and through sub-class, sibling class, equivalent class, disjoint class, to complete the classification of concepts.

In tea pest domain ontology, *tea pest, tea plant part, tea region, harm mode, harm degree, morphological characteristic, ecological control, pest natural enemy* are first defined as top-level classes, and then, subclasses of top-level classes are defined individually. For example, green leaf bug is a subclass of miridae, miridae and scutelleridae are sibling classes, and also class miridae and class scutelleridae are disjoint.

### 3.3    Property Definition

Ontology property contains object property and data property. Object property is used to associate instances and express the non-taxonomic relationships among the concepts of non-classified, with inverse, symmetry, transitivity and other constraints. The two associated classes are determined by its domain and range. Data property uses basic data types to describe characteristics of the instance, whose domain is used to define the property for a specific class, and the range including boolean, integer, character and the like.

This paper defines eat, eated, harm, harmDegree, harmMode, liveIn, pestControl, pestMorphology as object properties, at the same time, defines teaRegionCharacteristics, morphologyCharacteristics, lifeHabit, insectImage, harmSymptom, generationLife, environmentalFactor, controlMethods as data properties, and sub-properties have also been defined.

### 3.4    Instance Definition

Instance is a member of class. Creating instances need to adequately enumerate each member of the classes, and setting data properties and object properties in order to express the specific characteristics of instances. Moreover, ontology modeling requires a clear distinction between instance-of relationship and subclass-of relationship.

The instances of tea pest domain ontology are mostly concentrated in the classes of *tea pest* and *morphological characteristic*. Since the morphological characteristic of tea pest is so complex and various, some important discriminant signs of tea pest such as wing, wingspan, body length, body color must be refined and then defined as the instances of the *morphological characteristic* class when we integrate tea pest domain knowledge. For example, morphological characteristics of *green leaf bug-egg* include that *egg-length* is approximately 1.0 mm, *body-color* is yellowish green, etc.

### 3.5    Axiom Definition

Axiom has further elaborated on the concepts and relationships in order to make the logic of ontology more rigorous and contribute to the conflict detection and knowledge reasoning. Ontology axioms include class axioms, property axioms, instance axioms, constraint axioms and client defined rules.

The taxonomic relationships in ontology are defined and completed by class axioms, including *disjointClasses*, *subClassof* and *equivalentClasses*. Class axioms are defined while we create classes in the process of building ontology.

Property axioms include the restrictions of property domain and range, as well as symmetry, transitivity and other constraints. For example, the domain of *eat* is *pest natural enemy*, the range is *tea pest*, while the domain and range of *eated* is just the opposite and *eated* should be defined as the inverse property of *eat*.

Constraint axioms are divided into value restriction and cardinality restriction, which are mainly used to define the class of sufficient and necessary conditions. Instance axioms mean the declaration about ontology instances. While ontology axioms are defined through constraint on relationships either by graphic addition or rules description.

## 4    Reasoning and Evaluation on Tea Pest Domain Ontology

### 4.1    Ontology Reasoning

In order to check logic conflicts and get hiding information based on ontology, we choose FaCT++ to complete ontology reasoning. If there are any logic conflicts or case collisions, they should be amended according to the prompt and re-run FaCT++ till getting correct result.

There are several inconsistencies in tea pest domain ontology after reasoning and they have been revised according to the error messages. Running FaCT++ again and again, final testing result shows that tea pest domain ontology has a consistent logical structure and no instance conflicts are existed.

### 4.2    Ontology Evaluation

Ontology evaluation is essential to improve ontology quality, and is important foundation of the ontology construction. According to the content, it focuses on two aspects: concepts evaluation and relationships evaluation.

(1)Concepts evaluation: including concept completeness evaluation and concept accuracy evaluation. We test concepts accuracy and completeness of the constructing ontology by using a corpus-based concepts evaluation method. The method based on the comparison of the constructing ontology concepts and corpus terminologies, calculating the accuracy and recall rate of ontology concepts, accordingly, to complete effective concepts evaluation of tea pest domain ontology.

Accuracy is equal to the correct concepts divided by ontology concepts, and recall is equal to the correct concepts divided by concepts of the corpus. The measured accuracy and recall rate are shown in Table 1:

**Table 1.** Evaluation of the concepts in tea pest domain ontology

| Correct concepts | Ontology concepts | Concepts of the corpus | Accuracy | Recall |
|:---:|:---:|:---:|:---:|:---:|
| 493 | 564 | 625 | 87.50 % | 78.88 % |

(2)Relationships evaluation: including relations accuracy, consistency and simplicity evaluation. Aiming at relationships evaluation, this paper proposes an algorithm based on semantic similarity. The method firstly calculates the semantic similarity between two concepts, and semantic similarity is calculated as follows:

$$Sim(c_1,c_2) = \frac{2 \times depth(lso(c_1,c_2))}{len(c_1,c_2) + 2 \times depth(lso(c_1,c_2))} \tag{1}$$

depth(lso(c1,c2)): the depth of the common parent node between two concepts; len(c1,c2): the path length between two concepts.

According to the comparison results of the similarity value and threshold $\varepsilon$, if sim(c1,c2)$\geq\varepsilon$, then it conforms to the related property, otherwise it is inconformity and should points out the error types. Determine whether the relationships among concepts of the constructing ontology can fulfill consistency, accuracy, simplicity, thus we complete the relationships evaluation on tea pest domain ontology. The evaluation results are shown in Table 2:

**Table 2.** Evaluation of the relationships in tea pest domain ontology

| Relationships among concepts | Similarity | Consistency | Accuracy | Simplicity |
|:---:|:---:|:---:|:---:|:---:|
| Harm mode-Sapping | 0.724 | √ | √ | √ |
| Curculio chinensis-Curculio chinensis | 1.000 | √ | √ | Error |
| The southern China tea rgion-Taiwan | 0.875 | √ | √ | √ |
| Zeuzera coffeae-Xyleborus fornicatus Eichh | 0.093 | √ | Error | √ |
| ... | ... | ... | ... | ... |

In total there are 0 consistency errors, 7 accuracy errors, and 4 simplicity errors among relationships and all these errors have been corrected. According to the results of concepts evaluation and relationships evaluation, we make a comprehensive analysis of the constructed ontology, the conclusion is that tea pest domain ontology is well modeling, and achieves the goal of knowledge representation and also realizes the purpose of knowledge sharing and reuse.

## 5   Tea Pest Domain Ontology

Now, the tea pest domain ontology has 560 classes, 56 properties, 960 instances totally, including all the common tea pests in China and their relevant knowledge. Ontology can be graphically displayed by the plug-in OntoGraf, based on which the concepts and relationships are expressed intuitively and vividly.

The overall framework of tea pest domain ontology is shown in Figure 4, including all of the classes, properties and instances in this ontology, at the same time, concepts and relationships are clearly showed by the nodes and connections in this figure.

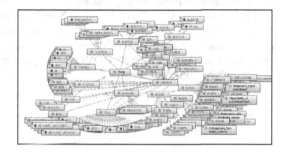

**Fig. 4.** Overall framework of tea pest domain ontology

We take tea pest domain ontology as knowledge base and build an ontology-based knowledge sharing platform which can provide the basis for ontology services such as diagnosis and control of tea pests, decision-making support for tea production, intelligent retrieval in tea pest domain. Additionally, apart from ontology-based knowledge services, we also need to persistently maintain and upgrade the ontology according to the knowledge renewal, only in this way can the better services be provided to customers.

## 6   Conclusion

Ontology, a tool of knowledge representation and knowledge reasoning, for ontology-based knowledge sharing, reuse and knowledge service, plays a fundamental role in the Semantic Web and knowledge engineering fields. In this paper, we firstly integrate the knowledge and information resource about tea pest; secondly, explore the procedures for the construction of tea pest domain ontology, and complete the development of the ontology; lastly, realize the sharing and reuse of tea pest domain knowledge.

However, time-consuming manual building of ontology is prone to generate bias errors. In order to improve the efficiency of ontology building, the next step is to use machine learning, natural language processing and other technologies to build ontology automatically or semi-automatically.

**Acknowledgement.** This work is supported by the National Natural Science Foundation of China (31271615, 30971691), and the University Provincial Science Foundation of Anhui (KJ2013Z077). Shaowen Li is the corresponding author of this paper.

# References

1. Studer, R., Benjamins, V.R., Fensel, D.: Knowledge Engineering, Principles and Methods. Data and Knowledge Engineering 25(1), 161–197 (1998)
2. Qian, P., Zheng, Y.: Agricultural Ontology Theory Research and Application. China Agricultural Science and Technology Press, Beijing (2006)
3. Sini, M., Salokhe, G., Pardy, C., et al.: Ontology-based Navigation of Bibliographic Metadata: Example from the Food, Nutrition and Agriculture. In: Proceedings of the International Conference on the Semantic Web and Digital Libraries, Rome, Italy, pp. 64–76 (2007)
4. Sini, M., Lauser, B., Salokhe, G., et al.: The AGROVOC concept server: rationale, goals and usage. Library Review 57(3), 200–212 (2008)

# Implementation and Evaluation of a Backtracking Algorithm for Finding All Justifications in OWL 2 EL

Zhangquan Zhou and Guilin Qi

Key Laboratory of Computer Network
and Information Integration of State Education Ministry
School of Computer Science and Engineering,
Southeast University, China
{quanzz,gqi}@seu.edu.cn

**Abstract.** Finding justifications plays an important role in the development and maintenance of ontologies. It helps users or developers to understand the entailments by presenting the minimal subsets of the ontology which are responsible for the entailment (called justification). In our previous work, we have proposed a backtracking algorithm for finding justifications in description logic $\mathcal{EL}^+$, which underpins OWL 2 EL, a profile of the latest version of Web Ontology Language (OWL). However, the algorithm is not implemented. In this paper, we implement this algorithm and conduct experiments on some real world ontologies, including Snomed-CT. We compare our algorithm with a black-box algorithm for finding all justification in OWL 2 EL. The experimental results show that in most cases, the backtracking algorithm outperforms the black-box one.

## 1 Introduction

The Web Ontology Language OWL has been designed as one of the major standards for formal knowledge representation and automated reasoning in the Semantic Web. The most recent version of OWL is called OWL 2. As OWL has begun to be used in many real world applications, many developers and users have participated in ontology building and editing. However, the development and maintenance of large ontologies are complex and error-prone. Some unwanted entailments may be inferred from an ontology containing errors. It is important to provide with users some explanation services to automatically find the erroneous axioms. Finding justifications is a reasoning service of this kind, which helps the users or developers to understand an entailment by presenting all minimal subsets (called justifications) of the ontology which are responsible for the entailment.

In recent years, there is an increasing interest in finding justifications. In general, the existing methods can be classified into two categories: glass-box approach and black- box approach. The glass-box approach is based on a reasoning algorithm of a description logic. There are some works that modify and

G. Qi et al. (Eds.): CSWS 2013, CCIS 406, pp. 235–238, 2013.

extend tableau-based algorithms for checking satisfiability of a DL ontology (see [7,6]). A glass-box method for $\mathcal{EL}^+$ is introduced in [2]. This method is based on the classification algorithm of $\mathcal{EL}^+$ [1] by attaching a label to each axiom in the original ontology and propagating it to inferred axioms during applying the completion rules.

The black-box approach treats a DL reasoner as an "oracle" and use it to check satisfiability of an ontology (see [2,3,5]). The black-box algorithms select the appropriate subsets of the ontology to the reasoner and check if these sets are the justifications for an entailment. The black-box approach can be optimized by module extraction techniques (see [3,8]).

In our previous work [4], we propose a backtracking algorithm for computing all the justifications for an entailment in $\mathcal{EL}^+$. Our method adapts the backtracking technique used by the glass-box approach for finding justifications. Unlike the glass-box approach, our method does not label entailments during inference but utilizes the classification result of the given ontology. However, the algorithm is not implemented. In this paper, we implement this algorithm by using multithreaded techniques and perform experiments on some real ontologies, such as Snomed-CT. We compare our algorithm with a black-box algorithm for finding all justification in OWL 2 EL. The experimental results show that in most cases, the backtracking algorithm outperforms the black-box one.

## 2    Implementation

In this section, we discuss the implementation of the backtracking algorithm proposed in [4] for finding justifications in $\mathcal{EL}^+$. We use a dependency graph to store the explanation sets and implement this algorithm. Our implemented program backtracks the dependance graph to find all justifications for the entailment. An *explanation set* is a set of axioms (including entailments) that can infer the entailment. The *dependency graph* can be defined as $\mathcal{G} = (V, E)$, where $V$ is the set of nodes and $E$ the set of edges. Each node $V_{\mathcal{E}}$ corresponds to an explanation set $\mathcal{E}$. For each edge $(V_{\mathcal{E}_1}, V_{\mathcal{E}_2})$, we have $\mathcal{E}_1 = \mathcal{E}_2 \cup \mathcal{E}_\alpha \setminus \{\alpha\}$, where $\mathcal{E}_\alpha$ is an explanation set of $\alpha$. Due to the page limit, we refer the interesting reader to our technical report [1] for the formal definitions and algorithms. We illustrate the notion of dependance graph using an example.

Suppose $\mathcal{O}$ is an $\mathcal{EL}^+$ ontology consisting of the following axioms:

$$\alpha_1 : A \sqsubseteq P_1 \quad \alpha_2 : A \sqsubseteq \exists r.P_2 \quad \alpha_3 : r \sqsubseteq s \quad \alpha_4 : P_2 \sqsubseteq P_1$$
$$\alpha_5 : \exists s.P_1 \sqsubseteq P_3 \quad \alpha_6 : P_1 \sqsubseteq B \quad \alpha_7 : P_3 \sqsubseteq B \quad \alpha_8 : \exists r.P_4 \sqsubseteq B$$

By applying the completion rules of $\mathcal{EL}^+$ [1], we have three entailments $A \sqsubseteq_{\mathcal{O}} B$, $A \sqsubseteq_{\mathcal{O}} P_3$ and $A \sqsubseteq_{\mathcal{O}} \exists s.P_2$ respectively denoted by $\beta_1$, $\beta_2$, $\beta_3$. For this example, we use the backtracking algorithm to compute all the justifications for $\beta_1$. See Figure 1, we use the classification result of the given ontology to backtrack the

---

[1] Our technical report can be downloaded from
   http://cse.seu.edu.cn/people/qgl/papers/csws-findj-EL.zip

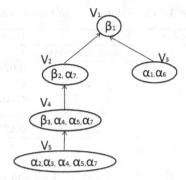

**Fig. 1.** The dependency graph for computing justifications for $A \sqsubseteq_{\mathcal{O}} B$

dependency graph in left part of Figure 1 but not label the entailments during reasoning, i.e., we backtrack from $\beta_1$ and check which sets of axioms (including inferred ones) can infer $\beta_1$ by applying one completion rule. In this case, $\{\alpha_1, \alpha_6\}$ and $\{\beta_2, \alpha_7\}$ are the two sets that satisfy this condition and they can infer $\beta_1$ by applying the completion rule CR1 [1]. Therefore two new nodes $V_2$ and $V_3$ are generated and added into the graph. Since $\beta_2$ is also an entailment, the backtracking continues from $\beta_2$ and $V_2$ is expanded to $V_4$. This process repeats until all the leaf nodes consist of axioms from the original ontology. These leaf nodes include all the justifications for the entailment. We can further process them to obtain all the justifications by removing the duplicates. Further, it is easy to implement this algorithm using multithreaded techniques based on this structure, because we can treat every node in the graph as an independent routine.

## 3    Experiments

We used five real ontologies to perform our experiments, namely Not-Galen, Galen, NCI, GO and Snomed-CT. Our running environment is an IBM X3850 server with a 16GB memory and 8 cores. To classify an ontology, we used **jcel** [2] as the $\mathcal{EL}^+$ reasoner. We compared our algorithm with the modularization-based black-box method given in [8]. Since there is no implementation of the modularization-based black-box method in java, we re-implemented this method by Java using **jcel** as the reasoner to do the subsumption checking.

For Not-Galen, NCI and GO, we sampled 20% of all the entailments that are well-distributed and representative of the ontology. For Galen and Snomed-CT, we sampled 200 subsumption entailments. In addition, since neither our algorithm nor the black-box one can finish justifications computing in an acceptable time for some cases, we set 10 minutes as the time limit for each test.

From Table 1, we can see that for Not-Galen, NCI and GO, our algorithm can finish for all samples in the limited time, while some cases run by black-box method reached time-out. For Galen and Snomed-CT, our algorithm reached time-out for

---

[2] http://jcel.sourceforge.net/

**Table 1.** The evaluation of two methods

| ontology | black-box method | | backtracking method | |
|---|---|---|---|---|
| | avg. time/s | pct. of time-out cases | avg. time/s | pct. of time-out cases |
| Not-galen | 0.216 | 0 | 0.024 | 0 |
| NCI | 1.918 | 2.5% | <0.01 | 0 |
| GO | 0.557 | 6% | <0.01 | 0 |
| Galen | 2.110 | 12.5% | 0.352 | 3% |
| Snoemd-CT | 0.732 | 36.5% | 0.421 | 30.5% |

some cases, but the proportion of such cases in all samples is less than that of the black-box method. Furthermore, without considering the time-out cases, the average time of computing justifications of our algorithm is less than that of black-box one, even without taking account of the module extraction time.

## 4   Conclusion

In this paper, We implemented the backtracking algorithm proposed in our previous work [4] for finding all justifications in $\mathcal{EL}^+$. Unlike the glass-box approach, our method is reasoner independent and does not attach labels to the entailments. We optimized it based on a new backtracking structure and implemented it by using Java multi-thread techniques. We compared our algorithm with an optimized black-box algorithm through evaluation on real ontologies. The experimental results showed that our algorithm performs better compared with the black-box algorithm in most cases.

## References

1. Baader, F., Brandt, S., Lutz, C.: Pushing the $\mathcal{EL}$ Envelope. In: IJCAI, pp. 364–369 (2005)
2. Baader, F., Peñaloza, R., Suntisrivaraporn, B.: Pinpointing in the Description Logic $\mathcal{EL}^+$. In: Hertzberg, J., Beetz, M., Englert, R. (eds.) KI 2007. LNCS (LNAI), vol. 4667, pp. 52–67. Springer, Heidelberg (2007)
3. Baader, F., Suntisrivaraporn, B.: Debugging SNOMED CT Using Axiom Pinpointing in the Description Logic $\mathcal{EL}^+$. In: KR-MED (2008)
4. Cheng, X., Qi, G.: An algorithm for axiom pinpointing in $\mathcal{EL}^+$ and its incremental variant. In: CIKM, pp. 2433–2436 (2011)
5. Kalyanpur, A., Parsia, B., Horridge, M., Sirin, E.: Finding All Justifications of OWL DL Entailments. In: Aberer, K., Choi, K.-S., Noy, N., Allemang, D., Lee, K.-I., Nixon, L.J.B., Golbeck, J., Mika, P., Maynard, D., Mizoguchi, R., Schreiber, G., Cudré-Mauroux, P. (eds.) ASWC 2007 and ISWC 2007. LNCS, vol. 4825, pp. 267–280. Springer, Heidelberg (2007)
6. Kalyanpur, A., Parsia, B., Sirin, E., Hendler, J.A.: Debugging unsatisfiable classes in OWL ontologies. J. Web Sem., 268–293 (2005)
7. Schlobach, S., Cornet, R.: Non-Standard Reasoning Services for the Debugging of Description Logic Terminologies. In: IJCAI, pp. 355–362 (2003)
8. Suntisrivaraporn, B.: Finding all justifications in SNOMED CT. Research Article, 79–90 (2013)

# $GORS$: A Graph-Based System for Ontology Revision

Yong Zhang*, Xuefeng Fu, and Guilin Qi

School of Computer Science and Engineering, Southeast University,
Nanjing 210096, China
{escmath,fxf,gqi}@seu.edu.cn

**Abstract.** In this paper, we present a Graph-based Ontology Revision System ($GORS$) for DL-Lite ontologies which implements a graph-based algorithm for ABox revision. $GORS$ transforms ontologies into graph structure and stores them in a Neo4j graph database. The graph-based algorithm does not compute the ABox closure w.r.t. the TBox, but it still takes into account of the logical consequences of the ontology. Thus, the system has potential to scale to ontologies with large ABoxes.

## 1 Introduction

Revising ontologies in Description Logics (DLs) deals with the problem of incorporating a new ontology into an old one consistently. This problem is important as DLs underpin W3C standard Web ontology language OWL and ontologies may evolve during their construction. Existing revision operators in DLs are mostly generalizations of belief revision operators in propositional logic.

Recently, there has been some work on revising ontologies in DL-Lite family, which is a family of DLs that provides tractable reasoning services. In [4,5], some model-based revision operators in DL-Lite are proposed. In [2], a plug-in system for ontology revision in Protégé, called OntoRevision, is given to implement the algorithm proposed in [5]. The system is based on a new semantic characterization called features, and aim to find maximal approximations in DL-Lite$_{bool}^{N}$. However, the efficiency of the system has not tested with large DL-Lite ontologies.

In [3], we defined a new operator for ABox revision in DL-Lite. A graph-based algorithm for ABox revision in DL-Lite is given to implement this revision operator. The advantage of this algorithm over the algorithms given in [4,5] is that the computational complexity of the graph-based algorithm is polynomial. Thus, it has potential to scale to ontologies with large ABoxes.

In this paper we describe an implementation of our graph-based algorithm and develop a system, called $GORS$, which focuses on dealing ABox inconsistency. We first introduce some basic notions of DL-Lite, and then introduce our algorithm for ABox revision. Finally, we present our system implementation and discuss what will be demonstrated.

---

* Corresponding author.

G. Qi et al. (Eds.): CSWS 2013, CCIS 406, pp. 239–242, 2013.
© Springer-Verlag Berlin Heidelberg 2013

## 2    Preliminaries

We start with the introduction of DL-Lite$_{core}$, which is the core language for the DL-Lite family [1]. The complex concepts and roles of DL-Lite$_{core}$ are defined as follows: (1) $B \longrightarrow A \mid \exists R$, (2) $R \longrightarrow P \mid P^-$, (3) $C \longrightarrow B \mid \neg B$, (4) $E \longrightarrow R \mid \neg R$, where $A$ denotes an atomic concept, $P$ an atomic role, $B$ a basic concept, and $C$ a general concept. A basic concept which can be either an atomic concept or a concept of the form $\exists R$, where $R$ denotes a basic role which can be either an atomic role or the inverse of an atomic role.

We extend DL-Lite$_{core}$ to DL-Lite$_{FR}$. In DL-Lite$_{FR}$, an ontology $\mathcal{O} = \langle \mathcal{T}, \mathcal{A} \rangle$ consists of a TBox $\mathcal{T}$ and an ABox $\mathcal{A}$, where $\mathcal{T}$ is a finite set of *concept or role inclusion assertions* of the form: $B \sqsubseteq C$; and $\mathcal{A}$ is a finite set of *membership assertions* of the form: $A(a)$, $P(a, b)$. To keep the logic tractable, whenever a role inclusion $R_1 \sqsubseteq R_2$ appears in $\mathcal{T}$, neither $(funct R_2)$ nor $(funct R_2^-)$ can appear in it.

We consider the revision of ontology $\mathcal{O} = \langle \mathcal{T}, \mathcal{A} \rangle$ by an ABox $\mathcal{A}'$. We assume that $\mathcal{O}$ is consistent and $\mathcal{T} \cup \mathcal{A}'$ is consistent, but $\mathcal{T} \cup \mathcal{A} \cup \mathcal{A}'$ is inconsistent. The problem of ABox revision is, how we modify $A$ such that $\mathcal{T} \cup \mathcal{A}'$ is consistent with the modified ABox.

## 3    Graph-Based Algorithm

Given a DL-Lite ontology $\mathcal{O}$ over a signature $\Sigma$, which can be partitioned into two disjoint signatures, $\Sigma_P$, containing symbols for atomic elements, i.e., atomic concept and atomic roles, and $\Sigma_C$, containing symbols for individuals. We briefly describe our method to construct a graph from a DL-Lite$_{FR}$ ontology.

According to the work given in [3], the digraph $G_\mathcal{O} = \langle N, E \rangle$ constructed from ontology $\mathcal{O} = \langle \mathcal{T}, \mathcal{A} \rangle$ over the signature $\Sigma$ is given as follows:

(1) for each atomic concept $B$ in $\Sigma_P$, $N$ contains the node $B$;
(2) for each atomic role $P$ in $\Sigma_P$, $N$ contains the node $P, P^-, \exists P, \exists P^-$;
(3) for each concept inclusion $B_1 \sqsubseteq B_2 \in T$, $E$ contains the arc $(B_1, B_2)$;
(4) for each role inclusion $P_1 \sqsubseteq P_2 \in T$, $E$ contains the arc $(P_1, P_2)$, arc $(P_1^-, P_2^-)$, arc $(\exists P_1, \exists P_2)$, arc $(\exists P_1^-, \exists P_2^-)$;
(5) for each individual $c$ in $\Sigma_C$, $N$ contains leaf node $c$;
(6) for each concept membership assertion $B(c) \in \mathcal{A}$, $E$ contains arc $(c, B)$;
(7) for each role membership assertion $P(a, b)$, $N$ contains node $(a, b), (b, a)$, and $E$ contains the arc $((a, b), P)$, arc $((b, a), P^-)$, arc $(a, \exists P)$, arc $(b, \exists P^-)$;

In our graph, each node represents a basic concept or a basic role, while each arc represents an inclusion assertion or a membership assertion, i.e. the start node of the arc corresponds to the left-hand side of the inclusion assertion (resp. individual or individual-pair of the membership assertion) and the end node of the arc corresponds to the right-hand side of the inclusion assertion (resp. concept or role of the membership assertion). In order to ensure that the information represented in the ontology is preserved by the graph, we add nodes $P, P^-, \exists P$,

$\exists P^-$ for each role $P$, $arc(P_1, P_2)$, $arc(P_1^-, P_2^-)$, $arc(\exists P_1, \exists P_2)$, $arc(\exists P_1^-, \exists P_2^-)$ for each role inclusion assertion $P_1 \sqsubseteq P_2$, and also $arc((a, b), P)$, $arc((b, a), P^-)$, $arc(a, \exists P)$, $arc(b, \exists P^-)$ for each role membership assertion $P(a, b)$ to the graph.

We now introduce algorithm *GraphRevi*, which takes $\mathcal{O} = \langle \mathcal{T}, \mathcal{A} \rangle$ and $\mathcal{A}'$ as its input. The algorithm can be explained as follows. Let $\mathcal{A}_{all} = \mathcal{A} \cup \mathcal{A}'$. It first computes the set $D$ of all the membership assertions in $\mathcal{A}$ that are in conflict with functionality axioms and $\mathcal{A}'$. It then constructs a digraph from $\langle \mathcal{T}, \mathcal{A}_{all} \backslash D \rangle$ and compute the set of all the membership assertions in $\mathcal{A}$ that are in conflict with some NI assertions and some assertions in $\mathcal{A}'$, and use this set to update $D$. These conflicts are caused by instances $a$ such that $(\mathcal{T} \cup \mathcal{A} \cup \mathcal{A}') \models \{X(a), \neg X(a)\}$. It means that if $v \in (Children(X) \cup Children(\neg X) \cup \{X\})$ and $v(a) \in \mathcal{A}$, then we can conclude that $v(a)$ results in $X \mapsto_G a$ or $\neg X \mapsto_G a$. We delete all such edges $\langle v, a \rangle$ in $G_{\langle \mathcal{T}, \mathcal{A}_{all} \backslash D \rangle}$. Meanwhile, all assertions, corresponding to these edges, in $\mathcal{A}$ are what we want to find. Thus, let $M = cl_\mathcal{T}(D) \backslash D$. The algorithm deletes all the membership assertions in $M$ that are in conflict with $NI$ assertions and $\mathcal{A}'$. Finally, $\mathcal{T} \cup (\mathcal{A} \backslash D) \cup M \cup \mathcal{A}'$ is the result of revision.

## 4   Implementation and Demonstration

We have implemented our graph-based algorithm for ABox revision in the Graph-based Ontology Revision System (*GORS*) in Java. *GORS* transforms a DL-Lite ontologies into graph structure and stores them in a Neo4j graph database, which is an open-source, high-performance and enterprise-grade *NOSQL* graph database. *GORS* provides the functionalities of revising ontologies in DL-Lite. In the user interface, users must select an ontology including TBox $\mathcal{T}$ and ABox $\mathcal{A}$, and ABox $\mathcal{A}'$ as input. Notice that $\mathcal{T} \cup \mathcal{A}$ and $\mathcal{T} \cup \mathcal{A}'$ are consistent respectively, but $\mathcal{T} \cup \mathcal{A} \cup \mathcal{A}'$ is inconsistent. Users can choose GraphRevi, using algorithm *GraphRevi* to revise them.

As shown in Figure 1, *GORS* has two major modules: (1) FileUpdate; (2) GraphRevi.

In the module of "FileUpdate", users can add files of the form owl or rdf to the system. Since the input of the algorithm is one TBox and two ABoxes, users should select appropriate files.

In the module of "GraphRevi", *GORS* provides five functionalities:

(1) **Visualization.** *GORS* can transform selected ontology into graph and present it in an intuitive way, as shown in the bottom of Figure 1.
(2) **ShowTree.** In *GORS*, TBox will be converted into a tree hierarchy so that users can have a clearer view of its structure (see left part of Figure 1).
(3) **Revision.** In this part, as mentioned above, users can revise $\mathcal{T} \cup \mathcal{A} \cup \mathcal{A}'$. After revision, users can obtain a revised graph and an OWL ontology file.
(4) **Closure.** *GORS* can compute closure for the TBox as well as for ABox w.r.t. TBox.
(5) **InconsistencyPath.** *GORS* also provides users with functionality of presenting inconsistency paths. That is, users can check those assertions causing inconsistencies and are deleted by *GORS*.

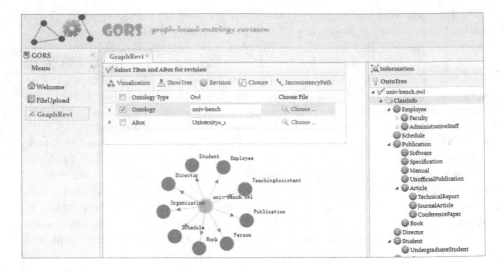

**Fig. 1.** The user interface of *GORS*

## 5   Conclusion

We have implemented a graph-based system for revising DL-Lite ontologies, called *GORS*. It is able to revise inconsistencies which is caused when incorporating an ABoxe to an ontology. The system is based on java and use Neo4j graph database to store ontologies. Apart from the system description given in this paper, we also evaluated the system over ontologies adapted from UOBM benchmark ontologies. Our preliminary experiments show that the system can handle DL-Lite ontologies with large ABoxes.

## References

1. Artale, A., Calvanese, D., Kontchakov, R., Zakharyaschev, M.: The dl-lite family and relations. J. Artif. Intell. Res. (JAIR) 36, 1–69 (2009)
2. Cobby, N., Wang, K., Wang, Z., Sotomayor, M.: Ontorevision: A plug-in system for ontology revision in protégé. In: JIST, pp. 417–424 (2011)
3. Gao, S., Qi, G., Wang, H.: A new operator for abox revision in DL-Lite. In: AAAI, pp. 2423–2424 (2012)
4. Qi, G., Du, J.: Model-based revision operators for terminologies in Description Logics. In: IJCAI, pp. 891–897 (2009)
5. Wang, Z., Wang, K., Topor, R.W.: A new approach to knowledge base revision in DL-Lite. In: AAAI, pp. 369–374 (2010)

# Integration of Pattern-Based Debugging Approach into RaDON

Qiu Ji[1,2], Zhiqiang Gao[1,2], and Zhisheng Huang[3]

[1] School of Computer Science and Engineering, Southeast University, Nanjing, China
[2] Key Laboratory of Computer Network and Information Integration
(Southeast University), Ministry of Education, Nanjing, China
{jiqiu,zqgao}@seu.edu.cn
[3] Department of Computer Science, Vrije University Amsterdam, the Netherlands
huang@cs.vu.nl

**Abstract.** RaDON is a representative system for debugging and repairing both single ontologies and ontology mappings. However, the efficiency of its debugging algorithms is still a problem. In this paper, we implement an efficient debugging algorithm based on patterns in RaDON. As a result, RaDON could efficiently discover those minimal unsatisfiability-preserving subsets (MUPS) with the patterns specified by the users. Besides, RaDON could discover all MUPS by reusing those MUPS that have previously been found.

## 1 Introduction

It is widely believed that ontologies play an important role for the formal representation of knowledge on the Semantic Web, especially when Web Ontology Language (OWL)[1] becomes a W3C recommendation standard. Description Logics (DLs), as the logical foundation of OWL, provide a well-defined semantics. Ontologies can be built either manually or semi-automatically by some learning algorithms. They can also be merged into larger ontologies through ontology mapping. Since building ontologies is error-prone, we often need to face the problem of dealing with logical inconsistency and incoherence[2] [3,6].

There are many algorithms for debugging DL-based ontologies, which can be classified into two approaches: a glass-box approach and a black-box approach. A glass-box approach is based on the reasoning algorithm of a DL [7]. In contrast, a black-box approach treats a DL reasoner as a "black-box" or an "oracle" and uses it to check satisfiability of an ontology [4]. A couple of systems have been developed to debug OWL ontologies. Among them, SWOOP[3] and RaDON[4] are two representative systems to debug an ontology. Apart from ontology debugging, RaDON can perform inconsistency-tolerant reasoning tasks and cope with incoherent ontology mappings. Therefore, it includes more functionalities than other systems for inconsistency handling.

---

[1] http://www.w3.org/TR/owl-ref/
[2] An ontology is inconsistent if it has no model. An ontology is incoherent if it contains at least one unsatisfiable concept which is interpreted as an empty set.
[3] http://code.google.com/p/swoop/
[4] http://atur.aturstudio.com/homepage/qiuji/radon.htm

G. Qi et al. (Eds.): CSWS 2013, CCIS 406, pp. 243–246, 2013.
© Springer-Verlag Berlin Heidelberg 2013

Although RaDON has implemented an optimized black-box debugging algorithm, it may not be able to deal with large ontologies [8]. To improve its efficiency, we implement an efficient algorithm based on some patterns proposed in [1] in RaDON. By doing this, RaDON could efficiently discover those MUPS with patterns specified by the users and all MUPS could be discovered by reusing those MUPS that have previously been found.

## 2    Background Knowledge

In this section, we briefly introduce the RaDON system presented in [2] and the pattern-based debugging approach given in [1].

### 2.1    The RaDON System

RaDON is a system to deal with inconsistency and incoherence for both ontologies and ontology mappings. It is originally developed as a plug-in for the NeOn toolkit[5] which is an extensible ontology engineering environment. RaDON[6] extends the capabilities of existing reasoners with the functionalities to deal with inconsistency and incoherence and mainly provides the following functionalities: 1) Debugging an ontology to compute all MUPS or minimal inconsistent subsets. 2) Repairing an ontology automatically by removing an axiom set satisfying minimal change. 3) Repairing an ontology manually based on the debugging results that have already been discovered. 4) Coping with inconsistency based on a paraconsistency-based algorithm [5].

### 2.2    Pattern-Based Debugging Approach

The pattern-based approach makes use of various heuristic strategies to compute MUPS instead of invoking a DL reasoner frequently. In fact, during its debugging process, only the functionality to minimize a set of supersets of MUPS needs to invoke a reasoner. Therefore, this approach could efficiently discover those MUPS with the patterns that have been implemented. Note that a MUPS may contain more than one pattern. In this work, the following four patterns have been implemented. Here, $X$, $Y$ and $Z$ indicate atomic concepts and $r$ is an object property whose domain and range are concepts.

**Subclassof-Disjoint Pattern.** It means that one concept $X$ is unsatisfiable if $X$ is a subconcept of two disjoint concepts. Namely, $X \sqsubseteq Y$, $X \sqsubseteq Z$ and $Y \sqcap Z = \bot$.

**Exist-Bottom Pattern.** It means that one concept $X$ is unsatisfiable if $X$ has an existential restriction whose filler is a bottom concept. Namely, $X \sqsubseteq \exists r.Y$ and $Y \sqsubseteq \bot$.

**Exist-All Pattern.** It means that one concept $X$ is unsatisfiable if $X$ has both existential restriction and universal restriction along the same property and the fillers of the two restrictions are disjoint. Namely, $X \sqsubseteq \exists r.Y$, $X \sqsubseteq \forall r.Z$ and $Y \sqcap Z = \bot$.

**Exist-Domain Pattern.** It means that one concept $X$ is unsatisfiable if $X$ has an existential restriction whose restricted property has a domain that is disjoint with $X$. Namely, $X \sqsubseteq \exists r.Y$, $\exists r.\top \sqsubseteq Z$ (i.e., $Z$ is a domain of $r$) and $X \sqcap Z = \bot$.

---

[5] http://www.neon-toolkit.org/

[6] For simplicity, RaDON in this paper actually means the RaDON plug-in in NeOn toolkit.

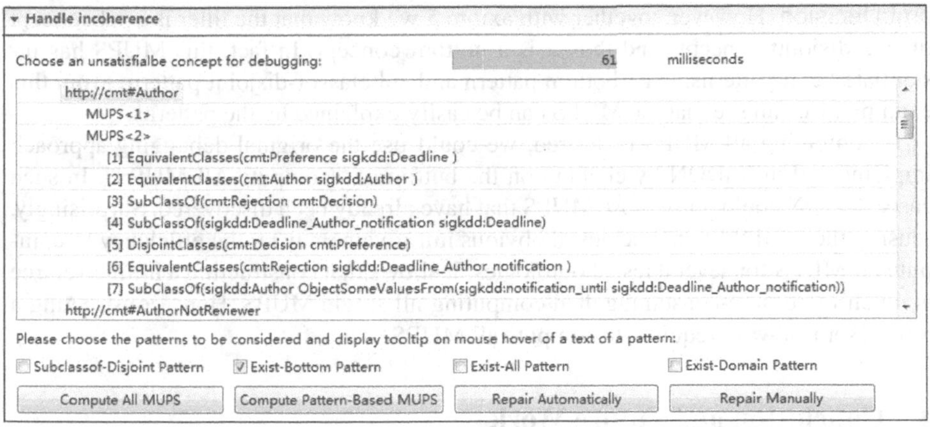

**Fig. 1.** The MUPS that are discovered according to the exist-bottom pattern

# 3 Computation of MUPS in RaDON

This section introduces the new debugging functionalities in RaDON by implementing the pattern-based debugging approach.

According to Figure 1, the users can choose one or more of the four patterns mentioned in Section 2. After that, those MUPS corresponding to the selected patterns can be discovered by clicking the button "Compute Pattern-based MUPS". This will provide the users with those MUPS with the same pattern and help them to understand the MUPS easier.

We demonstrate the system by considering the computation of MUPS for unsatisfiable concept cmt:Author in the ontology that is constructed by merging two ontologies cmt and sigkdd (the mapping between them is generated by the mapping system Maas-Match). The two ontologies and the mapping are provided by OAEI2012[7]. If we choose either "Subclassof-Disjoint Pattern" or "Exist-All Pattern", no MUPS can be discovered. When the "Exist-Bottom Pattern" is selected, two MUPS can be found within 61 milliseconds (see Figure 1). When the "Exist-Domain Pattern" is chosen, RaDON could discover 48 MUPS within 431 milliseconds.

To illustrate how we use the patterns to understand MUPS, we consider the MUPS displayed in Figure 1 which fall into the "Exist-Bottom Pattern" in Section 2. The debugged concept cmt:Author is unsatisfiable because it has an existential restriction whose filler is a bottom concept. According to axioms 2 and 7 (we use the number to indicate the axioms for simplicity), we know that cmt:Author has a restriction

$$\exists sigkdd:notification\_until.sigkdd:Deadline\_Author\_notification.$$

Next, we need to know why its filler sigkdd:Deadline_Author_notification is a bottom concept. From axioms 1 and 4, we infer that the filler is a subconcept of concept cmt:Preference. From axioms 3 and 6, we infer the filler is a subconcept of concept

---

[7] http://oaei.ontologymatching.org/2012/conference/

cmt:Decision. However, together with axiom 5 we know that the filler is a subconcept of two disjoint concepts and thus it is a bottom concept. In fact, this MUPS has instantiated two patterns: exist-bottom pattern and subclassof-disjoint pattern. From this example, we can see that the MUPS can be easily explained by the patterns.

If computing all MUPS is desired, we could use the original debugging approach implemented in RaDON by clicking on the button of "Compute All MUPS". In such case, RaDON could reuse those MUPS that have already been discovered. Surprisingly, reusing the MUPS has not achieved obvious improvement w.r.t. the efficiency to compute all MUPS for several tested ontologies. It shows that constructing a hitting set tree is much more time-consuming than computing all single MUPS. Here, constructing a hitting set is always required to compute all MUPS.

## 4    Conclusion and Future Work

We have implemented the pattern-based debugging approach in RaDON to improve its efficiency. Furthermore, all MUPS could be discovered by reusing those MUPS that have already been found by patterns. We demonstrated that RaDON can quickly discover those MUPS corresponding to the patterns specified by the users and these MUPS can be easily understood by checking the patterns.

As a future work, we intend to provide a graphical user interface for facilitating users to understand the MUPS according to the patterns specified by users.

## References

1. Ji, Q., Gao, Z., Huang, Z., Zhu, M.: An efficient approach to debugging ontologies based on patterns. In: JIST, pp. 425–433 (2011)
2. Ji, Q., Haase, P., Qi, G., Hitzler, P., Stadtmüller, S.: RaDON — repair and diagnosis in ontology networks. In: Aroyo, L., Traverso, P., Ciravegna, F., Cimiano, P., Heath, T., Hyvönen, E., Mizoguchi, R., Oren, E., Sabou, M., Simperl, E. (eds.) ESWC 2009. LNCS, vol. 5554, pp. 863–867. Springer, Heidelberg (2009)
3. Kalyanpur, A., Parsia, B., Sirin, E., Cuenca-Grau, B.: Repairing unsatisfiable concepts in OWL ontologies. In: Sure, Y., Domingue, J. (eds.) ESWC 2006. LNCS, vol. 4011, pp. 170–184. Springer, Heidelberg (2006)
4. Kalyanpur, A., Parsia, B., Sirin, E., Hendler, J.: Debugging unsatisfiable classes in OWL ontologies 3(4), 268–293 (2005)
5. Ma, Y., Hitzler, P., Lin, Z.: Algorithms for paraconsistent reasoning with OWL. In: Franconi, E., Kifer, M., May, W. (eds.) ESWC 2007. LNCS, vol. 4519, pp. 399–413. Springer, Heidelberg (2007)
6. Meilicke, C., Stuckenschmidt, H., Tamilin, A.: Repairing ontology mappings. In: AAAI, pp. 1408–1413 (2007)
7. Schlobach, S., Cornet, R.: Non-standard reasoning services for the debugging of description logic terminologies. In: IJCAI, pp. 355–362 (2003)
8. Stuckenschmidt, H.: Debugging OWL ontologies - a reality check. In: EON (2008)

# Author Index